I0450015

The Dueling Loops of the Political Powerplace

WHY PROGRESSIVES ARE STYMIED AND
HOW THEY CAN FIND THEIR WAY AGAIN

A Thwink.org Project

Many thanks to the hundreds of people who have contributed to evolving this book. The potential of what you hold in your hands should be ample proof your efforts were worthwhile.

Let's hope that with this book (and similar efforts) social problem solvers can move from treating the symptoms to resolving root causes, and from changing politicians to changing the system. [1]

Happy reading and welcome to a whole new way of thwinking....

Thwink.org
1164 DeLeon Court
Clarkston, GA 30021 USA
jack@thwink.org
Jack Harich

Third Edition
Printed by Lulu Press
ISBN 978-1-4303-2973-2
Copyright © 2014 by Thwink.org

Cover photograph by the author on October 3, 2002 from a ridgeline near the top of La Dent Parrachee in Parc National de la Vanoise, France.

Contents

4

Scope and Message

THE DUELING LOOPS ARE A MODEL OF A KEY PORTION OF THE POLITICAL POWERPLACE. The scope of this book is limited to a hypothesis of what the structure of this model looks like, how it behaves, and how that knowledge can be applied. If you find yourself thirsting for further discussion of the many other concepts briefly introduced here, please see the additional material at Thwink.org.

This book carries three main messages. The first is the Dueling Loops model explains the mystery of why progressives have been unable to reliably solve difficult social problems. *This is the diagnosis.* The model also predicts how, if progressives switched to pushing on high leverage points instead of low leverage ones, they could solve such problems with relative ease. *This is the treatment.* As modern medicine has demonstrated, successful treatment requires correct diagnosis.

Underneath lies a more subtle second message. *The Dueling Loops model is an example of how, once activists can clearly "see" the dynamic structure of the problems they are working on, what to do to solve them will become relatively obvious.* Like the astronomer who now has a telescope, activists will be able to see and do a multitude of marvelous things that were beyond their reach before. Solving difficult social problems will still require serious work. But it will now be a much more pleasurable and successful experience, because the human system will respond in a more predictable manner.

Going even deeper, activists will develop sound models and use them to solve difficult problems only if they are driven by *a process that fits the problem.* This is the third and deepest message. Activists are problem solvers. A process that fits the problem will become *the* foundation for progressivism, just as the process of double entry accounting became the foundation for the business world in the 15th century and the Scientific Method became the foundation for all of science in the 17th century.

What's the difference between a good problem solver and a great one? I believe it's the ability to ask the right question at each fork in the road as a problem is solved. If you have a process that fits the problem, the process automatically guides you toward what those questions should be at the strategic level. As Toyota says, "The right process will produce the right results." [2]

That the process must fit the problem is *the* message of this book.

Introduction

T HIS IS NOT A BOOK ON HOW TO SOLVE THE SUSTAINABILITY
PROBLEM. Society's proven inability to do that can only be a symptom
of a deeper problem: its inability to solve most difficult social problems, including war, poverty, corruption, excessive inequality, and environmental
sustainability. All these problems have defied solution for thousands of years.
WHY IS THIS?

The approach this book takes to answer that question is so uncommon that
when I presented early versions of the concept to environmentalists, the reaction of nearly all of them was to reject it outright. Even career professionals,
some with MBAs or PhDs and one CEO earning $150,000 a year, rejected it.

This left me with the baffling problem of how to best express my ideas.
How do you take a concept that goes 180 degrees against the norm, and hence
is almost certain to be rejected, and communicate it in such a manner that
automatic rejection does not occur?

My answer was to go over and over the core of the argument until it was
so simple it was easily understood and so logically compelling it was immediately accepted as self-evident, once you understand it. This is the basic structure of the Dueling Loops, as shown on the cover of this book. This almost
perfectly symmetrical shape answers what to one group of activists is the
toughest question in the world:

> The goal of progressives is to promote the common good for all. In
> theory this is also the goal of democracy. *Why then do democratic
> systems so strongly resist changing their behavior from what benefits
> the special interest few to what benefits the common good of all?*

In other words, why is change resistance so strong, when it comes to solving progressive problems whose solution would so obviously benefit the
common good? The system should welcome such change, but it's doing just
the opposite by benefitting special interests. This is the real problem to solve
and can be called the **Progressive Paradox**.

Stunning proof that democratic systems are biased toward benefiting special interests instead of the common good appeared in 2014 in a paper by
Gilens and Page, titled *Testing Theories of American Politics: Elites, Interest
Groups, and Average Citizens*. Using entirely new data compiled by the authors in the study, the paper concluded that "economic elites and organized
groups representing business interests have substantial independent impacts
on U.S. government policy, while mass-based interest groups and average

citizens have little or no independent influence." While the study focused on the U.S., the results appear to be generalizable to any political system where "economic elites and organized groups representing business interests" play a major role.

So how do we answer the question posed by the Progressive Paradox: *Why do democratic systems so strongly resist changing their behavior from what benefits the special interest few to what benefits the common good of all?*

Our answer and our departure from the norm begins with this line of reasoning:

Turning our Attention to the Social Side of the Problem

Most effort on solving the sustainability problem focuses on its **technical side**, which consists of the proper practices (technologies or behaviors) that must be followed in order to achieve sustainability. Examples of proper practices are renewable energy, permaculture, and the four R's of reduce, reuse, recycle, and repair. *But surprisingly little effort addresses why most of society resists adopting these practices.* This is the change resistance or **social side** of the problem.

Change resistance is the tendency for a system to resist change even when a surprisingly large amount of force is applied, in an attempt to solve a problem. The problem can be any difficult social problem, such as sustainability, poverty, war, corruption, innumerable types of discrimination and exploitation, or the perennial tendency for the gap between the rich and those beneath them to grow.

The Dueling Loops answer the question that is rapidly becoming *the* question of the 21st century: Why is civilization unable to solve the sustainability problem in time? Or as we have more properly framed the question: Why is change resistance to solving the sustainability problem so strong?

What is it about the answer that runs so counter to conventional wisdom? Let me try to explain, using the most important problem of them all, sustainability, as an example. (As you read the rest of this book, remember that the sustainability problem is a proxy for all problems whose solution would benefit the common good.)

In 1972 the *Limits to Growth* project and book conclusively identified the global environmental sustainability problem. Ever since then, millions of environmentalists, ranging from grassroots activists all the way up to those working with international efforts like the United Nations Environmental Programme, have been furiously trying to solve the problem. But they have failed. While there has been some success on easy problems like local pollution, the more difficult problems like climate change, deforestation, top soil loss, and freshwater scarcity remain as unsolved as ever. Why is this? Why is the system so strongly resisting change?

Another example of change resistance occurred in 1999 when the United States Senate voted 95 to zero against signing the Kyoto Protocol. Not a single senator could be persuaded to vote for the world's best hope of solving the climate change problem, even though a democratic president (Bill Clinton) and a rather pro-environmental vice president (Al Gore) were in office at the time.

The technical versus the social side of the problem is a crucial distinction. Society is aware of the proper practices required to live sustainably. But most of society has a strong aversion to adopting these practices. As a result, problem solvers have created thousands of effective (and often ingenious) proper practices, but they are stymied in their attempts to have them taken up by enough of the population to solve the problem. *Therefore the social side is the crux of the problem and must be solved first.* [3]

But that is not what environmentalists are doing.

Instead, in every case I've examined so far, environmentalists are mostly trying to solve the technical side of the problem. I have yet to find a single individual or organization focusing on the social side, though there must be some. *This shows problem solvers have been working on solving the wrong problem, which is a striking conclusion that should send shockwaves throughout all of environmentalism.*

Consider the old saying, "You can lead a horse to water but you can't make him drink." Problem solvers have been working on finding the water (finding technical solutions) or leading the horse to it (promoting those solutions and putting them under the horse's nose). But that's the easy part. What they should be working on instead is *how to get the horse to decide to drink.* [4]

Strategy

In mid 2001, after 20 years as a consultant, I made helping to solve the sustainability problem my life's work and committed to the project full time. As a systems engineer from Georgia Tech, my specialties have been small business management, process improvement, problem analysis, information and software engineering, and all sorts of related topics.

When I started the project I immediately set up a six year, three step strategic plan. The first two years were for getting my arms around the problem. The next two were for making an original contribution. The last two were for communicating my ideas and starting to work elbow to elbow with others to combine my possibly useful ideas with theirs to solve the environmental sustainability problem. This is when the first edition of the Dueling Loops book was written.

On top of this three step plan I imposed two key strategies. The first was to work in isolation for the first four years. This was because no significant progress had been made, indicating a novel approach was needed. But if I worked with others or based my research on the literature *instead of the actual system*, then I would probably fall into the same ruts and groupthink traps as others. Hence the critical importance of working alone at first. The drawback to self-imposed isolation is lack of networking and remaining an unknown in fields you are trying to influence. Normally this is a surefire road to failure. It was a tough choice, but I was prepared to take that chance.

The second strategy was far more important. From day one I set about designing a formal process to solve the problem. This became the System Improvement Process. What separates it from the rest is decomposition of the sustainability problem into three distinct subproblems. The first is overcoming change resistance. This is the strategy that led to discovery of the Dueling Loops of the Political Powerplace.

The Progressive Side of the Book Is Born

The purpose of building the Dueling Loops model was to answer the question in step 2.1 of the System Improvement Process, as listed on page 169: *Why is there such strong resistance to adopting the solution?* As so often happens in scientific explorations, a pleasant surprise occurred. Although I was addressing the sustainability problem, the model turned out to be so generic that it also explains why there is such strong resistance to adopting a solution *to any difficult progressive problem.*

This was a tremendous insight. But what to do with it? Fortunately the perfect opportunity appeared when I realized that the *Analytical Activism* book, at a ponderous 262,000 words, was simply too big and serious for most readers. The solution was to extract what interested readers the most and put it into a much smaller book (which has about 70,000 words). This was the analysis of the Dueling Loops model. When I begin to design the little book, I could see this was a chance to frame the model differently. *Instead of a model for the change resistance part of the sustainability problem, I elevated it up one level of abstraction to be a model for the fundamental challenge all progressives face: how to get political systems to accept their new viewpoints, ones that would benefit the common good.*

This was exciting because I could see the potential. The Dueling Loops really do seem to explain the phenomenon of systemic change resistance. As you work your way through the book, you will see this is basically because progressives are working for the good of the system as a whole. Their goal is to optimize the system for the common good of all, rather than the good of the special interest few, which is the opposition's goal. The Dueling Loops explain how these two opposing goals are basically two opposing feedback loops in the political system. Whichever loop gains the most supporters wins. Currently the wrong loop is dominant, which is why progressives are so stymied, frustrated, and helpless, because they have no idea this is the cause of decades of problem solving failure.

By reframing the Dueling Loops as an analysis of the Progressive Paradox, this book aims to help not just one but two very worthy types of readers: frustrated environmentalists and equally frustrated progressives. This book offers a strategic path out of that agony. The path consists of using three key tools: root cause analysis, a problem solving process that fits the problem, and modeling the problem. The book illustrates how to apply these tools by using the sustainability problem as a running example.

The Contents of the Book

Part 1: Getting Started, frames the problem by describing The Progressive Paradox. The real problem to solve is progressivism has long been blocked from achieving its ideals, due to systemic change resistance. But why? If we can solve that mystery we can overcome the resistance. The system will then change from resisting solutions to naturally "wanting" to solve progressive problems, starting with the most important one of them all: sustainability.

The foundation of how to crack the problem wide open begins with this carefully worded definition:

> *Progressive philosophy is a comprehensive rationale and value set whose goal is optimizing the human system for the common good of all and their descendents.*

This becomes the catalytic concept that I hope will carry you through the rest of the book, just as it carries me forward in my own work.

After framing the problem, Part One then introduces the tools that will be used to solve it. The hypothesis that change resistance is the crux of the problem is presented.

Part 2: The Dueling Loops Model and Sample Solution, is the intellectual meat of the book. It presents the Dueling Loops model and six sample solution elements that push on the high leverage point found in the model. Also presented is the New Dominant Life Form (the modern corporation and its allies) and the five main types of political deception. Using the same computer simulation approach that *The Limits to Growth* used, a series of 22 model scenarios are explored. By comparison the first edition of *The Limits to Growth* used 12 scenarios. Just as in *The Limits to Growth*, it is the description of the model and these scenarios that are the heart of the book, *because they explain so much and, if true, allow us to use the model to begin to solve what up until now have been insolvable problems.*

Part 3: The Niche Succession Model and Sample Solution, is short. It extends the Dueling Loops by adding the Niche Succession subsystem. This explains what's really happening at the deep level Darwin would be thinking on if he was alive and working on the problem today. An ecological niche succession is underway. The Previous Dominant Life Form, *Homo sapiens*, has been surpassed by the New Dominant Life Form, who is now in control of the biggest niche on the planet: the biosphere. The extended model reveals another high leverage point: quality of political decision making. The solution element of Decision Ratings is presented to push on this point. Decision Ratings promise to radically improve the effectiveness of political systems, just as the invention of modern democracy did 200 years ago.

Part 4: How Can We Apply This New Knowledge? The book answers this question in three unique ways:

Chapter 11: The Assault on Reason Examined, moves from theory to practical application with an educational critique of Al Gore's book, *The Assault on Reason*. The chapter shows where he went somewhat astray in his search for "trying to figure out what has gone wrong in our democracy, and how we can fix it" and how he could correct that error, using the perspective of the Dueling Loops and true analysis. I have tried to be very diplomatic and sensitive here. The helpful critique applies to all similar books, articles, and efforts, a point I hope that you and other readers will see.

Chapter 12: Taking Up Where Limits to Growth Left Off, proposes a project taking up where *The Limits to Growth* left off in 1972. The premise is that *The Limits to Growth* only identified the sustainability problem. Now we need to take the next step. This is not to solve it, as conventional wisdom assumes. Instead, the next step is to diagnose why the system is so strongly resisting changing to a sustainable mode. Once a correct diagnosis is made, then we can go ahead with developing a solution. History shows this will be an order of magnitude easier to implement than those being attempted now, because we have at last diagnosed why the patient is ill.

Chapter 13: The Tantalizing Potential of a Permanent Race to the Top, finishes on the highest note possible by exploring the prospect of a permanent race to the top. The difference between this vision and others is it's based on a structured analysis of how to make this state come about. *This is realistically possible and even probable once the Dueling Loops are understood.* This is a vision people can rationally get excited about, because it arises from a comprehensive, experimentally provable analysis. To me this leads to *rational optimism* instead of *emotional optimism*. There is a difference.

Going Deep

Consider this book's historic context. *The Limits to Growth* used a simulation model to correctly identify the sustainability problem. No other tool could have done that. Due to the extreme difficulty and complexity of the sustainability problem, the same tool is required to take the next step, or it will fail.

But there's more. What the Dueling Loops book is doing at the deepest strategic level is executing a process that fits the problem. This is the System Improvement Process, a generic process for solving any difficult social problem. As the Scope and Message page concludes, "That the process must fit the problem is *the* message of this book." Lack of a process that fits the problem is the ultimate reason progressives are stymied, no matter what country they may live in or what problem they are working on.

I sincerely hope that after you've finished reading the book, these points ring loud and clear and true, because if they do, then we can solve the Progressive Paradox.

Part One

Getting Started

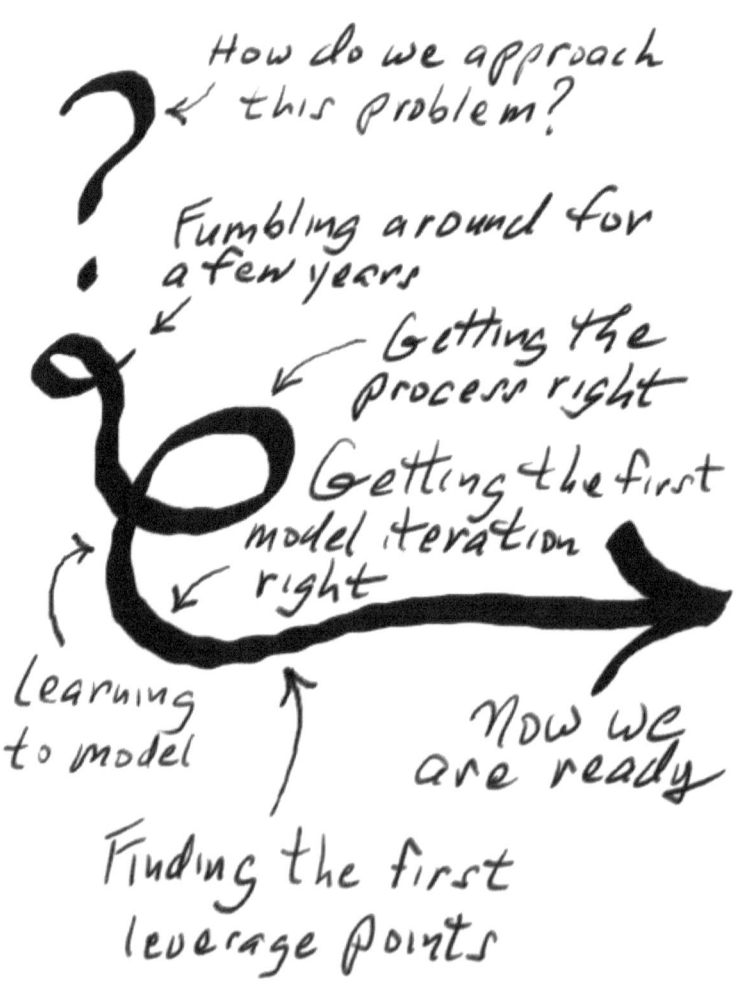

Chapter 1

The Progressive Paradox

C IVILIZATION HAS BEEN TRAPPED IN A GRIM PARADOX FOR FAR TOO LONG. In a world of plenty, too many have too little. In what could be a world of peace, one conflict after another is the norm. In a world that could be brimming with honesty and virtue, corruption is far too common, even in developed countries. In a world that could be environmentally sustainable, local ecological collapse has occurred countless times, and is about to happen again, though this time on a global scale.

Why do these problems occur again and again, with no end in sight? How can this predicament be resolved?

The predicament *is that progressivism has long been blocked from achieving its ideals, due to systemic change resistance.* This gives us two more terms to define: progressivism and change resistance.

On the surface, progressivism is a political movement and philosophy that supports causes like peace, worker's rights, social justice, control of the excesses of corporatism, and environmental sustainability.[5] It is a worldwide movement, because these problems are endemic to all cultures. The paradox is that solving all of these problems is physically possible and desirable, but so much system pushback occurs that solutions are at best partial and temporary.

That is just the superficial definition, however. Going deeper, let's define **progressive philosophy** as:

> *A comprehensive rationale and value set whose goal is optimizing the human system for the common good of all and their descendents.*[6]

This definition encompasses all the problems mentioned above, as well as many more. It follows that **degenerate philosophy** is just the opposite: a comprehensive rationale and value set for optimizing the system for the good of the few (the special interests), who are the degenerates. These two definitions allow us to see that ever since the beginning of politics, political systems have exhibited a pattern of behavior that is related to all the above problems: *Sometimes the degenerates are in control and sometimes they are not.* But even when they are mostly out of office they still retain such control of the system that their influence is pervasive and never ending. The result is too many problems dear to progressives are never fully or permanently resolved.

A **degenerate** is someone who has fallen from the norm. They have degenerated. The race to the bottom loop presented later explains why this occurs so easily. The term is not meant as a demeaning label, but rather as a hopefully temporary fall from virtue.

These definitions, progressive and degenerate, form our point of departure. They establish the premise that all the rest of this book builds upon. If you believe these are not the two fundamental ends of the modern political strategy spectrum in democracies, when all else is stripped away, then this is not the book for you. Or if you are a member of a special interest group that you believe is performing such a beneficial, indispensable service that it is entitled to special treatment, then this is not the book you should be reading. Or if you believe that democracy was not invented with the main goal of optimizing the common good of all, and that the real goal of democracy is or should be something else, then you should stop reading right here.

Earlier we observed that progressivism has been blocked from achieving its ideals due to *systemic change resistance*. Let's define that term. At the social system level, **change resistance** is the tendency for an entire system to resist change even when a surprisingly large amount of force is applied. At the individual social agent level, change resistance is the refusal of a person or organization to fully support or adopt new behavior. When we speak of systemic change resistance it is the first definition we are using. **Systemic** means affecting an entire system, as opposed to a small portion of it.

Several readers have commented that change resistance arises from people's values and that until you change those values, overcoming change resistance is impossible. But from a systems thinking point of view, values are not systemic. They are local. They are a symptom, an outcome, of something deeper that is occurring at the system level.

We can now state the paradox that forms the core of the problem this book seeks to solve:

The Progressive Paradox

1. Most people are progressives.
2. The goal of progressive philosophy is to promote the common good.
3. In theory this is also the goal of democracy.
4. **Why then do democratic systems so strongly resist changing their behavior from what benefits the special interest few to what benefits the common good of all?**

Serving the common good of all instead of the few is exactly the problem the invention of democracy was designed to solve. But it has not. Therefore something in the current model of democracy is flawed. Something deep inside present forms of democracy is causing governments to resist change that is for the common good. It's as if a deadly worm has gnawed its way to the very roots of the tree of democracy, and continually threatens to topple the tree, all because the system is resisting changing to behavior that's good for the tree. Therefore, until the root cause of systemic change resistance is found and resolved, no amount of hard work, inspirational appeal, or political maneuvering will proactively solve difficult progressive problems.

Finding that root cause is the most important thing this book will attempt to do. But before we begin, let's pause to more precisely define a term.

A Few Words about a Word

Our careful definition of progressive philosophy led immediately to its polar opposite: degenerate philosophy. In the word "degenerate" lies the potential for the message of this book to be misinterpreted, so let's correct any misconceptions now.

The term **degenerate** should not be taken as pejorative. It is not meant to demean or vilify. Instead, it is merely a label for those who have fallen into the clutches of the race to the bottom. They have lost their way. For most this is a temporary condition and can be rectified.

According to the *Random House Unabridged Dictionary*, the verb degenerate means "to fall below a normal or desirable level in physical, mental, or moral qualities." Senator Hayakawa, in *Choose the Right Word*, says "degenerate is relative, implying a decent from a higher state or better condition. It may indicate moral, physical, or mental deterioration from a standard or norm." [7]

With this issue cleared up, let's take the first step toward solving the paradox. As in most perilous journeys, the first few steps make the greatest difference.

Chapter 2

Extracting Ourselves from the Progressive Paradox

W E HAVE IDENTIFIED THE PROBLEM AND DEFINED OUR KEY TERMS. NEXT, HOW CAN WE BEST GO ABOUT extracting civilization from the Progressive Paradox?

Already a few clues are beginning to emerge. If you examine the basic problems that continually confront progressives, like poverty in a world of plenty, discrimination, war, corruption, and environmental sustainability, you will notice that in each case, *someone is benefiting from the problem.* All of these problems have existed for thousands of years or more. The severity of the problems seem to come and go, often with long steady rises capped by sudden falls, as the system seems to undergo some sort of endless mixture of cycles, of war and peace, of corruption and virtue, of excessive concentration of wealth and then dispersal, and so on.

Of all these problems, the worst of the worst is government corruption, because once corruption at the top begins, the system is broken. It is no longer running for the good of the people. Instead, it is run for the good of the few. Furthermore, because the degenerates want as much as they can get and they want it now, other problems receive less than the priority they deserve, such as the biggest one of them all: global environmental sustainability.

These patterns of behavior are a symptom that something in the system is broken. Something must be causing these symptoms. They are too consistent to be caused by mere chance.

We have the clues. We have the patterns. We know that strong systemic change resistance exists. But what does all this mean?

Until recently it was impossible to deeply and correctly answer that question, because it was not until the 1950s that one of the three tools we need was invented. This was system dynamics.

System Dynamics

System dynamics is a computer simulation modeling tool. Its purpose is to more deeply and correctly understand the dynamic behavior of social systems. System dynamic models emphasize the feedback loops of systems, using stocks, auxiliary variables, and flows of influence. System dynamics uses a standard visual notation and an interrelated collection of mathematical equations to mimic a system's important structure, *with the goal of gaining new insights into how and why the system works the way it does.* A computer program then runs the equations, which simulates the behavior of the system. The chief output is graphs showing the dynamic behavior of the system under the assumptions used for each particular simulation run. Below is a typical system dynamics modeling tool in action.

This is the user interface of Vensim. The version of Vensim used to produce the models in this book is free.

Vensim is as easy to use as a spreadsheet and only a little more difficult than a word processor. The user first "draws" the structure of a social system using stocks (the rectangles), auxiliary variables, and arrows. Next the simple mathematical equations for each node (the stocks

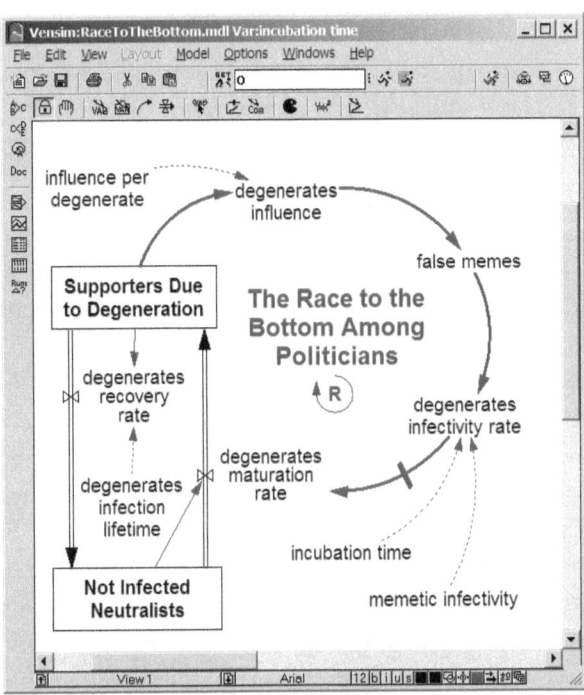

and auxiliary variables) are entered. For example, the equation for the <u>degenerates influence</u> node is <u>Supporters Due to Degeneration</u> times <u>influence per degenerate</u>. The equations are simple. It is the emergent behavior of the system the model represents that is complex.

Then the model is run and its behavior is examined via the use of graphs showing how the values of key nodes change during the simulation run. This knowledge is used to iteratively improve the model until its design objectives are achieved, such as finding the root cause of change resistance.

The world needs more modelers. Almost anyone who can use a computer, loves technical stuff, and has a strong interest can learn how to do computer simulation modeling. All that is required to learn the basics is to spend about a day with someone who knows how to model. Then inspect several good models and figure out what makes them so good. Next create a few practice models on your own, and a few more, until you start attaining your design goals.

After that new insights should start tumbling out faster than you know what to do with them. Sooner or later you may want to study the best book there is on this type of modeling: John Sterman's *Business Dynamics: Systems Thinking and Modeling for a Complex World.*

Eventually you may reach the same conclusion that I and many others have, that:

> *The supreme advantage of system dynamics modeling is the way it allows you to capture all your important assumptions about how a system behaves. It then allows you to accurately simulate the emergent behavior of that system, no matter how complex the model becomes.*

This is a bona fide miracle. The unaided human mind (or even 500 minds) cannot come anywhere close to doing this, except for the simplest of systems. System dynamics is *the* tool for identifying, analyzing, and solving difficult social system problems at the tactical level. (The right process is *the* tool at the strategic level.)

To the few that took the time to learn the tool, the impact of the invention of system dynamics in the late 1950s was as momentous as the telescope and microscope, because now social problem solvers could see something they had never seen before: *the structure of social systems.* Suddenly, in a few isolated pockets of science, problem solvers were able to make dramatic progress where they had been blocked before. An outstanding example was the work of Professor Jay Forrester of MIT, who was also the inventor of system dynamics. He was working on a case of the "In a world of plenty, too many have too little" problem. In less than one year's time he was able to solve the toughest problem of them all in the United States: the urban decay crisis.

In the 1950s and 1960s, urban decay and the symptoms it caused was America's biggest problem. It would eventually reach the crisis stage with the Los Angeles race riot of 1965, which left 34 people dead. Other riots occurred in Newark and Detroit. The problem continued to deteriorate, and in 1968 Martin Luther King Jr. was assassinated, which sparked further riots, including some in the nation's capitol. The riots, high levels of crime, growing discrimination and race hatred, and a host of factors increased white flight from inner cities. Businesses also moved out. This made the urban decay problem even worse, causing a vicious cycle. Despite a plethora of attempted solutions, the problem failed to get better. By the late 1960s the situation looked hopeless.

Into this void stepped Professor Jay Forrester. After a long and thorough examination of the problem, Forrester constructed a simulation model that conclusively demonstrated that four of the US government's top solutions ranged from no effect to making the problem worse. *None were making the urban decay problem any better.*

The four solutions were job programs for the underemployed, training programs for the underemployed, financial aid to cities for welfare and education, and low cost housing construction. Forrester's model showed why this last solution element turned out to be the worst of them all.

One reason was low cost housing attracted poor, low skilled people to the city. But the main reason was it preempted the use of the land the housing was on for other types of construction, such as housing for the middle and upper class, new businesses, and business growth. This so disrupted the needs of the majority of people and businesses in the city that in the simulation model, new enterprise fell by 49%, mature business fell by 45%, slums increased by 45%, and taxes rose 36%. While unemployment was down 4%, overall the outcome was a disaster. This agreed with what was occurring in the real world. Low cost housing was the most popular of all the solutions at the time, and paradoxically had the worst effect.

Forrester then proceeded to stun the cozy little world of urban management with a second even more astonishing discovery. Buried in the model were several *high leverage points* (defined later in this chapter) that no one had ever tried, because they were so invisible and counterintuitive. But when he ran the model and pushed on the high leverage points with hypothetical solutions, the symptoms of urban decay disappeared.

The overall solution employed a combination of policy changes designed to reverse the decay seen in the model. This included new enterprise construction, declining industry demolition, slum housing demolition, discouraging housing construction, encouraging industry, and an end to the four solutions mentioned above. Forrester's model demonstrated how intuitively derived solutions would not work. Due to the complexity of the problem, only solutions based on a thorough understanding of the dynamics of the problem would work, with experimentation and fine tuning as necessary.

These results were so startling and in such direct conflict with common sense and conventional wisdom that Forrester's work was at first ridiculed and attacked. But when he presented the model and the reasoning behind it in a series of five hour educational sessions, most participants accepted the conclusions and took up the cause. It was not long before Forrester's solutions and others suggested by this new way of thinking were tried. They worked. Today urban decay is still a problem, but it is no longer a crisis. The downward spiral of urban decay has largely been resolved.

A Process that Fits the Problem

How did Forrester so ingeniously apply the powerful tool of system dynamics to the urban decay problem? Did he rely on intuition and the fact that he was the inventor of the tool, and thus knew it so well he could use it to solve problems as easily as you and I use a browser to surf the web?

No. Behind that tool was another even more powerful one: a process that fit the problem. Over the years Forrester had developed a process that allowed him to quickly size up a business or social problem, model what mattered, and use the insights gleaned from the model to solve the problem.

We have done the same for difficult social problems. This is the System Improvement Process, created from scratch solely for this class of problems. It fits the sustainability problem and other progressive problems so well that the effort and skill required to solve them falls by an order of magnitude. This changes difficult problems from impossible to solve to solvable, just as Forrester's process did for the urban decay crisis. *All that is required is to conscientiously apply the process (which includes continuous process improvement) and that the problem be difficult but not impossible.* The details of the System Improvement Process are presented later on page 169.

The key to successful process driven problem solving is continuous process improvement. For truly difficult problems, the problem solving process itself is the deciding limitation. This is especially true for problems requiring a long period to solve and problems of an evolving nature. By applying the principle of continuous process improvement and letting the process drive your work, the process can be incrementally improved until it's good enough to solve the problem. Examples are the Toyota Production System, Kaizen, the scientific community's continuous improvement of the Scientific Method over many centuries, and the business community's continuous improvement of financial planning/accounting, which began in earnest with invention of double entry accounting in the thirteenth century.

Root Cause Analysis

At the heart of the System Improvement Process lies the most powerful tool of them all: root cause analysis. This is widely used in business and science. Examples are Six Sigma and NASA's Root Cause Analysis Tool. A **root cause** is the deepest cause in a causal structure that can be resolved. If it can't be resolved it's not a real problem. It's the way things are.

All problems require root cause analysis to solve, whether the term is used or not, since all problems arise from their root causes. In easy problems the root causes are obvious, so formal root cause analysis is not required. But in difficult problems, ones that have resisted solution repeatedly, the root causes are not obvious. They are so well hidden and so often counter intuitive that root cause analysis or its equivalent is required. There is simply no other way.

Insights and High Leverage Points

How the three tools work together is explained in the diagram. All three tools share the same goal: to lead problem solvers to the insights necessary to solve difficult social problems, ones so difficult they defy conventional approaches. This is a book about how to use these tools. *It is a book of how to find and apply insights, rather than an idea cookbook.* Cookbooks don't work, because you would need a different solution recipe (an idea) for every problem. Instead, this book teaches how to solve difficult social problems, by introducing you to how to use the two main tools to find solutions yourself.

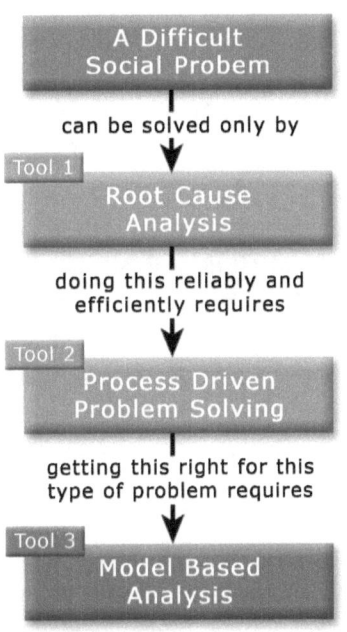

An **insight** is profound knowledge reflecting the inner nature of something, such as $E = MC^2$. Ideas are more superficial. An **idea** is knowledge reflecting the outer nature of something, like the way a pocketknife (or these days, a multi-purpose cell phone) is a handy tool to carry. The relationship between the two is captured in the title of Phil Dusenberry's 2005 book: *A Great Insight Is Worth a Thousand Good Ideas.* In the book he shows over and over how "a good insight can fuel a thousand ideas" and how good insights endure "because [their] basic truth [does] not change over the years." The application of insights like Einstein's famous equation has led to millions of new ideas. But ideas don't breed many more ideas. At most they lead to a few. *Insights lead to ideas, and ideas lead to what to do.*

Even though this is a book of insights, not ideas, it is strewn with hundreds of glittering ideas. Do not be led astray. The ideas are not what's important. Each idea is an example of how to apply a greater insight. The solution path the book presents is thus paved with illustrative examples. If you sit yourself down on a mountaintop and read and reread the book from a strategic perspective, *you will see its real purpose is to explain how to use the right tools to create the right insights that lead to the right ideas needed to solve problems.*

For example, the System Improvement Process (SIP) is an insight. ***SIP is so insightful it's an insight generator.*** Among other things, SIP helps you to find high leverage points. A high leverage point is NOT a place where a small change makes a big difference. That standard definition ignores how much effort it takes to make the change, such as the way Donella Meadows, in *Leverage Points: Places to Intervene in a System*, defines leverage points as "places within a complex system where a small shift in one thing can produce big changes in everything." [8] What a small shift might be is never defined.

Our definition of a **high leverage point** (HLP) is a place in a system where a small amount of change force (the total effort required to prepare and make a change) causes a large amount of predictable, favorable response. In the System Improvement Process, root causes are resolved by pushing on their related HLPs with various solution elements. An HLP is not a solution. ***An HLP is an insight into a solution strategy.*** In difficult social problems that take years to solve and whose solution consists of dozens to tens of thousands of tweaks to the system, there is no exact solution. There is only a solution strategy, which is pushing on the right HLPs.

That is ultimately all this book is about—how to find the right HLPs. Once they are found, the problem is 80% solved.

Applying the Right Tools

What might happen if system dynamics and a process that fit the problem were applied to the paradox that has bedeviled civilization since the dawn of history?

The question has been asked. The tools have been applied. The first major discovery was that the real problem is not what we think it is. Instead, we have reached an interesting hypothesis, one that sees change resistance as the crux of the sustainability problem.

Chapter 3

Change Resistance as the Crux of the Problem

THE PURPOSE OF THE DUELING LOOPS MODEL IS TO PERFORM STEP 2.1 OF THE SYSTEM IMPROVEMENT PROCESS: *Why is there such strong resistance to adopting the solution?* To answer this question we must first introduce a new term, proper coupling, so that we can more clearly understand change resistance.

Proper coupling occurs when the behavior of one system affects the behavior of other systems in a desirable manner, using the appropriate feedback loops, so the systems work together in harmony in accordance with design objectives. For example if you never got hungry you would starve to death. You would be improperly coupled to the world around you. In the environmental sustainability problem the human system is improperly coupled to the greater system it lives within, the environment.

The old paradigm: Proper coupling as *the problem* to solve

The universal consensus is that how to achieve proper coupling is *the* problem to solve. The early literature of global sustainability framed the debate this way.

In 1972 *The Limits to Growth* brought the environmental sustainability problem to the world's attention. The book defined the problem as how "to establish a condition of ecological and economic stability that is sustainable far into the future." [9] In other words, how can we properly couple the ecological and economic systems, by finding and implementing the right policies to keep environmental impact at a sustainable level? Works like *The Limits to Growth* and its predecessors, notably Rachel Carson's *Silent Spring* in 1962 and Jay Forrester's *World Dynamics* in 1971, firmly established what can be called "proper coupling" as *the* problem to solve.

Subsequent analyses and dialog strengthened this perspective into the dominant paradigm. In 1987 the United Nations' Brundtland Report stated that "*Our Common Future* serves notice that the time has come for a marriage of economy and ecology...." [10] In 1997 the nascent field of ecological economics argued that "three policies to achieve sustainability" are "a broad natural capital depletion tax, application of the precautionary polluter pays principle, and a system of ecological tariffs." [11] These are proper coupling mechanisms. They attempt to internalize externalized costs, which itself is a proper coupling perspective. In 2007 an IPCC report stated that: "A wide variety of policies and instruments are available to governments to create the incentives for mitigation action. They include integrating climate policies in wider development policies, regulations and standards, taxes and charges, tradable permits, financial incentives, voluntary agreements, information instruments, and research, development and demonstration." [12] These too are proper coupling mechanisms.

Because proper coupling is seen as *the* problem to solve, finding and implementing the right coupling policies has become the *raison d'être* of the sustainability movement. But if we examine the problem from another perspective and decompose it differently, it's possible to take a much more productive approach, one that is driven by:

The new paradigm: Change resistance as *the real problem to solve*

Years ago I was discussing a perplexing problem with Steve Alexander, a bright young engineer/manager from the UK. He suggested that if you've looked at a problem from all angles and are still stumped, then you probably have a *missing abstraction.* Find it and the difficulties will melt away.

Change resistance is that missing abstraction.

Change resistance is the tendency for a system to resist change even when a surprisingly large amount of force is applied. Difficult social problems are best decomposed into two sequential subproblems: *How to overcome change resistance* and *How to achieve proper coupling.* This is the timeless strategy of divide and conquer. By cleaving one big problem into two the problem becomes much easier to solve, because we can approach the two subproblems differently and much more appropriately.

There's a simple reason this decomposition works so well: change resistance is usually what makes difficult social problems difficult. In fact, regardless of whether change resistance is high or low, it is impossible to solve the proper coupling part of a complex system social problem without first solving the change resistance part. This is nothing new. As the old joke goes, "How many psychologists does it take to change a light bulb? Just one. But first the light bulb has to want to change."

In difficult social problems the system spends a long time trying to overcome change resistance. Once that occurs proper coupling is achieved relatively quickly by introduction of new norms, laws, and related mechanisms, and is refined still further over time. This pattern has occurred in countless historic social problems whose solution benefits the common good, like universal suffrage, slavery, racial discrimination, the rule of colonies by other countries, the recurring war in Europe problem (solved by creating the European Union, which properly coupled member nations together to reduce pressures for future wars), and the self-perpetuating ruler problem (solved by invention of democracy, which properly coupled the people and their rulers via the voter feedback loop). True to form, the pattern is occurring again in the sustainability problem.

Here's what the third edition of *Limits to Growth* had to say about change resistance. The term was never used, because it was a missing abstraction. Note the final sentence, which says it all: (Italics added)

"[The second edition of *Limits to Growth*] was published in 1992, the year of the global summit on environment and development in Rio de Janeiro. The advent of the summit seemed to prove that global society had decided to deal seriously with the important environmental problems. But we now know that humanity failed to achieve the goals of Rio. The Rio plus 10 conference in Johannesburg in 2002 produced even less; it was almost paralyzed by a variety of ideological and economic disputes, [due to] the efforts of those pursuing their narrow national, corporate, or individual self-interests.

"...*humanity has largely squandered the past 30 years...* " [13]

And here's what Jorgen Randers, co-author of all three editions of *Limits to Growth,* had to say recently. Again, note the missing abstraction: (Italics added)

"This brings us back to the starting point of [Jay Forrester's] original analyses. The early world models recommended 'equilibrium.' They prescribed a limited rate of investment, only enough to replace depreciation; a rate of births limited to only replace deaths; resource use less than we can get from technological advance; and pollution less than the absorptive capacity of the globe. These are good *recommendations*, and probably unavoidable recommendations. But *they were politically infeasible*—both in 1970 and 2000—and they will possibly remain so for a long time.

"In 1970 system dynamics defined the overshoot problem, and described *the sustainability solution.* In 2000 sustainability is still far off, in spite of the early warning the world dynamics studies gave. This simply goes to demonstrate a well known truth: System Dynamics is powerful—*the challenge lies in implementation,* or in biblical language: 'the mind is willing, but *the flesh is weak.'* " [14]

Because of the missing abstraction "the sustainability solution" does not apply to the total sustainability problem. It applies only to the proper coupling subproblem, which the passage treats as *the* problem to solve. That's why it goes through the trouble of listing proper coupling solutions in the form of "recommendations." The charges that "they were politically infeasible... the challenge lies in implementation... the flesh is weak" acknowledge the *real* problem: that change resistance has been so high it has thwarted 30 years of efforts to achieve proper coupling. The result, as the first passage lamented, is that "humanity has largely squandered the past 30 years."

What is the underlying cause of such prolonged, pervasive change resistance? Whatever it is, it must be incredibly strong to cause such a powerful effect.

In business management change resistance has long been known as resistance to change, organizational momentum, or inertia. Peter Senge's business classic, *The Fifth Discipline*, describes the structural cause of organizational change resistance this way: (Italics added)

> "In general, balancing loops are more difficult to see than reinforcing loops because it often looks like nothing is happening. There's no dramatic growth of sales and marketing expenditures, or nuclear arms, or lily pads. Instead, the balancing process maintains the status quo, even when all participants want change. The feeling, as Lewis Carroll's Queen of Hearts put it, of needing 'all the running you can do to keep in the same place' is a clue that a balancing loop may exist nearby.

> "Leaders who attempt organizational change often find themselves unwittingly caught in balancing processes. To the leaders, it looks as though their efforts are clashing with sudden resistance that seems to come from nowhere. In fact, as my friend found when he tried to reduce burnout, the resistance is a response by the system, trying to maintain *an implicit system goal. Until this goal is recognized, the change effort is doomed to failure.*" [15]

This applies to the sustainability problem. Until the "implicit system goal" causing systemic change resistance is found and resolved, change efforts to solve the proper coupling part of the sustainability problem are, as Senge argues, "doomed to failure."

Three premise argument that change resistance is the crux

The transformation of society to environmental sustainability requires three steps: The first is the profound realization we must make the change, because if we don't our descendants will suffer immensely, due to environmental and economic collapse. The second is finding the proper practices that will allow living sustainably. The third step is adopting those practices.

Society has faltered on the third step. (1) By now the world is aware it should live sustainably, which is the first step. (2) There are countless practical, proven ways to do this (or the gap can be easily closed), which is the *proper coupling* or technical side of the problem and the second step. (3) But for strange and mysterious reasons, society has not yet taken the final step to adopt these practices, which is the *change resistance* or social side of the problem. *Therefore change resistance is the crux of the problem.*

Let's examine the evidence for the three premises in the above paragraph:

Premise 1: The world is aware it should live sustainably

First published in 1972, *The Limits to Growth* became an international best seller and went on to sell thirty million copies. Many early adopters became aware the world must shift into a sustainable mode.

The message took time to spread, but finally in 1992 the United Nations Conference on Environment and Development was held in Rio de Janeiro. 172 governments attended, with 108 sending their heads of state. There were 9,000 journalists, 35,000 activists, politicians, and business representatives, countless casual attendees, and 25,000 troops to keep order. "Known as the Earth Summit, this was the largest environmental conference in history; in fact, it was probably the largest non-religious meeting ever held." [16]

The Rio Summit made the world's leaders aware we must take the message of *The Limits to Growth* seriously and start living sustainably. To show their commitment to this consensus the summit resulted in five documents: the Rio Declaration on Environment and Development, Agenda 21, Convention on Biological Diversity, Forest Principles, and Framework Convention on Climate Change. The last was the forerunner of the Kyoto Protocol.

"Agenda 21 is a comprehensive plan of action to be taken globally, nationally and locally by organizations of the United Nations System, Governments, and Major Groups in every area in which human impacts on the environment." [17] While it has not met its goals, the existence of Agenda 21 and the fact it has been signed by 178 governments shows that, in general, the world is fully aware it should live sustainably.

Premise 2: The proper practices already exist or are easily found

Proper practices are those behaviors necessary to achieve proper coupling. For the sustainability problem, examples of proper practices are smaller families, converting to renewable energy, and the four Rs of reduce, reuse, recycle, and repair. Looking at China, other examples of effective proper practices that already exist are a one-child per family policy and the use of 545 million bicycles versus only 7 million cars.

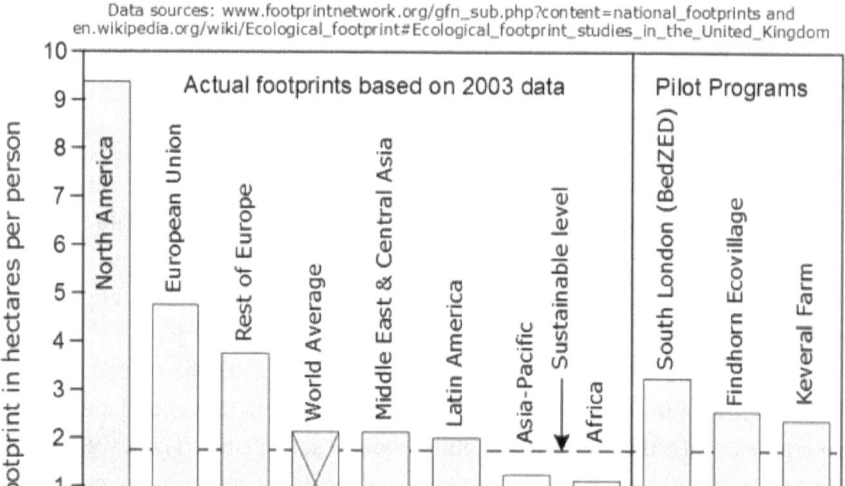

Ecological Footprints   Graph created by Thwink.org

Data sources: www.footprintnetwork.org/gfn_sub.php?content=national_footprints and en.wikipedia.org/wiki/Ecological_footprint#Ecological_footprint_studies_in_the_United_Kingdom

Consider the data above. [18] Several ecological footprint pilot programs in the UK have demonstrated that we already have the proper practices needed to reduce the footprint from an average of 5.45 global hectares per capita (gha) in the UK to levels of 3.2, 2.56 and 2.4 gha, *while maintaining comfortable standards of living.* [19] Considering that a footprint of 1.8 gha is needed to be sustainable, we are already almost there with easy to deploy off-the-shelf practices. The remaining gap is easily closed by further research and experimentation. There are some deficiencies with ecological footprint measurement, but overall, the pilot programs demonstrate this premise is probably true.

Premise 3: Society has not adopted the proper practices

Society has made small gains in reducing environmental impact. But these have only been enough to slow impact growth, not bring it down to a sustainable level, as illustrated below. [20]

Each of the five dots on the curve was a major event in the course of environmentalism. The first two dots, Silent Spring and Limits to Growth, brought the sustainability problem to the world's attention. The Brundtland Report, which famously rede-fined sustainability as sustainable development,

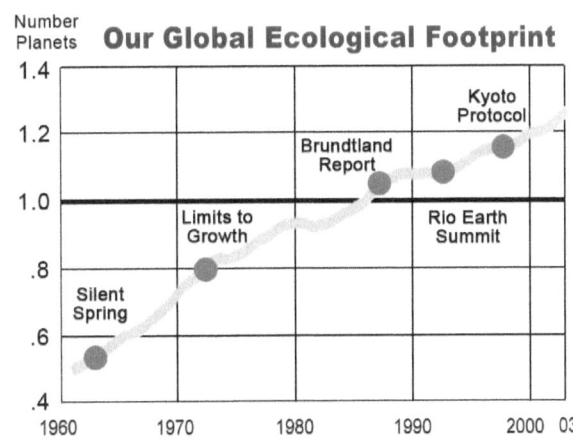

and then defined that as "development that meets the needs of the present without compromising the ability of future generations to meet their own needs," attempted to start the world on a path to a solution. The last two dots are international efforts and commitments that are part of that solution.

But change resistance is so high that none of these events had more than a negligible impact on the growth of the ecological footprint. The curve marches steadily upwards, unstoppable as an elephant, as vivid proof that society has not yet changed to the proper practices necessary to bring the foot-print down to the one planet line.

Conclusion

These three premises all appear to be true. *It follows that change resis-tance is the crux of the problem.* Change resistance *must* be overcome first. Then, and only then, does the proper coupling problem become solvable. But this has not happened, because environmentalists see proper coupling as *the* problem to solve and hence have been trying to solve it first. *As a result, civi-lization has spent 30 years trying to solve the wrong problem.*

This opens up a major new line of attack. First society solves the change resistance part of the problem. Once that's done the system will "want" to be properly coupled, because it's already aware it should live sustainably. Proper coupling will then occur surprisingly quickly because most of the proper prac-tices are already known.

Solving the change resistance part of the problem would cause a **phase transition**. This occurs when a system moves (or sometimes jumps) from one mode to another, due to having crossed a critical threshold. Here the threshold is the *temporary* amount of force (best applied at a high leverage point) it takes to overcome change resistance. In the pre-transition phase change resistance is significant. Just as water disappears when it changes to ice, in the post-transition phase change resistance has vanished or is insignificant. It is replaced by a strong tendency for the system to seek the new equilibrium of proper coupling. If the system is well understood the phase transition can be made to happen quickly, predictably, and with a minimal amount of force.

Thwink.org has produced a two hour *film* called *Cracking the Mystery of the Progressive Paradox*. The film visually explains the vital importance of overcoming change resistance with this image:

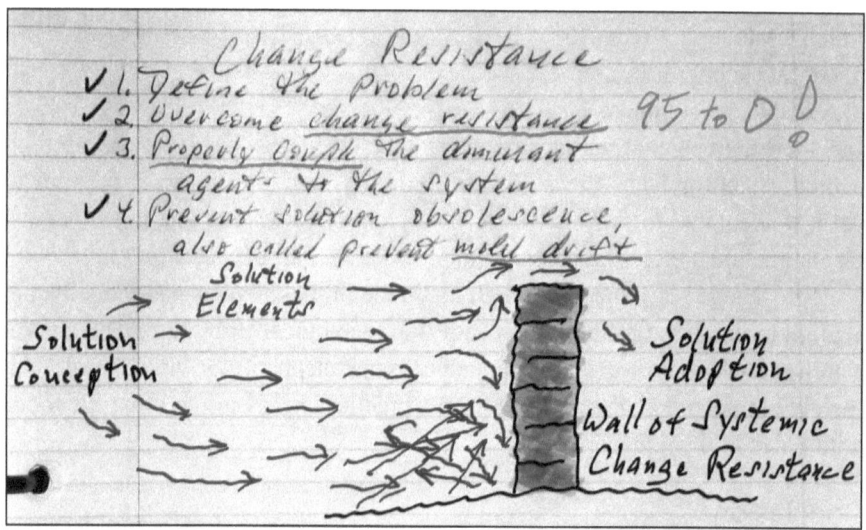

The film makes the point that until we can tear down that wall of change resistance, only a few proposed solutions will make it to solution adoption. The rest will hit the wall, bounce off, and accumulate at the bottom of the wall, in a veritable graveyard of worthy but rejected solutions. Over time these have piled up. So has the frustration of environmentalists.

But as this book argues, there is a better way.

Objection: But we're already working on change resistance!

Some readers may disagree and insist that environmentalists *are* trying to overcome change resistance. That's what environmental magazines, lobbying to get politicians to support sustainability, promotion of the proper practices, and so on are for, they say. If we can get enough people to accept the truth of these ideas, then society will start adopting the proper practices needed to live sustainably, and soon the problem will be solved.

But that's not really solving the change resistance problem. It's attempting to solve the proper coupling problem through endless communication and education about "the truth." It's like saying over and over to a four year old who won't drink his milk, "Please drink this. It's good for you. Let me tell you one hundred and one reasons why if you drink this glass of milk, you will grow up to be big and strong . . . ."

Can you see how this will fail if change resistance is high? So let's take a different approach: *What is the underlying cause of why Johnny won't drink his milk?* That's where we start treating this as a change resistance problem.

Suppose the cause is that unlike most children, Johnny hasn't outgrown milk allergy and doesn't like milk. This affects about 2% to 3% of infants worldwide. Most children outgrow milk allergy by age 2 or 3.[21] But Johnny hasn't, much to the consternation of his parents, who run a dairy farm. Everyone on the farm drinks milk. It's just the way things are.

Now suppose most of the four year olds on nearby dairy farms have the same problem. Then we have a *systemic* problem. There's something else happening besides a simple milk allergy. We obviously need to find the cause. Otherwise there's no way we are going to be able to solve the problem.

This is what we mean by *high systemic change resistance*. When it's present the proper coupling part of a problem cannot be solved first. Instead, you have to roll up your sleeves and first solve the change resistance part of the problem.

In the sustainability problem, instead of why Johnny won't adopt the practice of drinking milk, we have the problem of why politicians, corporations, and people won't adopt the practice of living sustainably. This has been the case for over 30 years. *Why is that?* Some areas of the world are way out in front, like the European Union. *Why is that?* A large amount of resistance comes from large for-profit corporations, who have been stalling, blocking, and lobbying against stricter regulations. *Why is that?*

As soon as we start thinking like this and putting all these questions together, we have switched from solving the proper coupling problem to solving the change resistance problem.

Two questions

The realization that we've spent 30 years trying to solve the wrong problem raises two questions. One looks behind, while the other looks ahead.

The first question is this: *Why have we squandered 30 years on solving the wrong problem?* To those who have worked on hundreds of significant problems of many kinds over their career and have frequently used a process to solve the difficult ones, the strategic answer is obvious: It's because problem solvers used a process that did not fit the problem. They used a traditional, intuitive, ad hoc process that had no conception of change resistance, social system analysis, root causes, and low and high leverage points. This works fine for easy problems, where change resistance is low. But it usually fails or takes a very long time on difficult problems, where change resistance is high, because problem solvers end up pushing on intuitively attractive but low leverage points.

Activists are a tiny minority, so they lack the force needed to make pushing on low leverage points work. Trying to turn an aircraft carrier around with a crowbar doesn't work—unless you can find the right high leverage points.

The second question takes much longer to answer: *If overcoming change resistance is the crux of the sustainability problem, then how can society solve the change resistance problem?*

That is what the rest of this book is all about.

A Related Paper

In 2010 Thwink.org published a paper titled *Change Resistance as the Crux of the Environmental Sustainability Problem.* This paper uses an entirely different approach (and simulation model) to analyze change resistance and accordingly arrives at very different insights from this book. Readers interested in a deeper understanding of the phenomenon of systemic change resistance and how it applies to public interest activism are urged to consider reading the paper.

Part Two

The Dueling Loops Model and Sample Solution

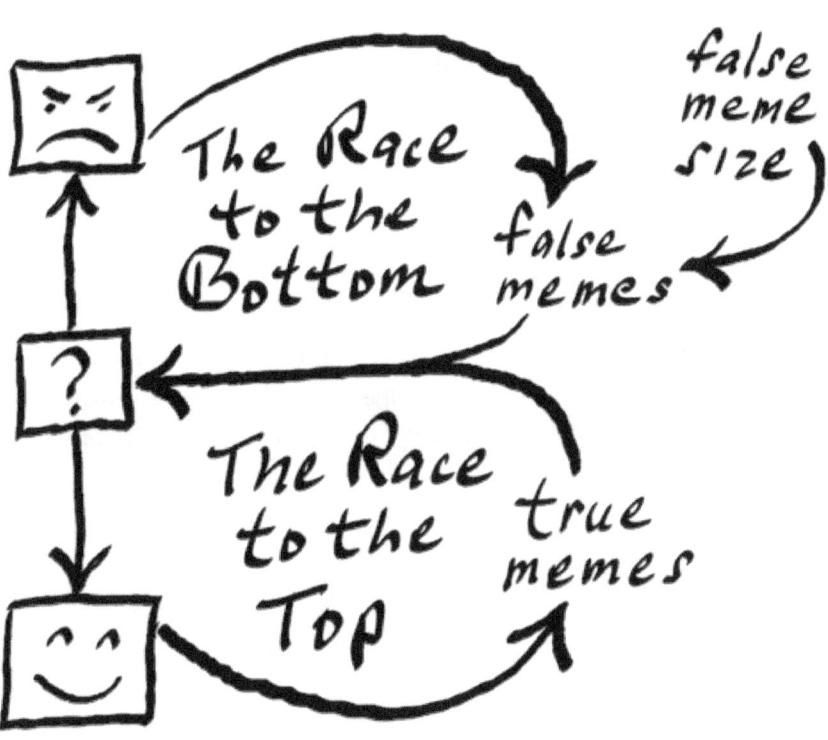

Chapter 4

The Basic Dueling Loops

THE MOST IMPORTANT PROBLEM FACING THE WORLD AND PROGRESSIVES IS THE SUSTAINABILITY PROBLEM. Using that problem as an example, here's an overview of what the Dueling Loops are all about:

Most effort on solving the sustainability problem focuses on its **technical side**, which is the proper practices that must be followed to be sustainable. But surprisingly little effort addresses why most of society is so strenuously resisting adopting those practices, which is the *change resistance* or **social side**.

Our analysis of the social side of the problem employs a relatively simple simulation model. The model shows the main source of change resistance lies in a fundamental structure called The Dueling Loops of the Political Powerplace. This consists of a race to the bottom among politicians battling against a race to the top. *Due to the inherent structural advantage of the race to the bottom it is the dominant loop most of the time, as it is now.* As long as it remains dominant, resistance to living sustainably will remain high.

The analysis has, however, uncovered a tantalizing nugget of good news. There are promising *high leverage points* in this structure that have never been tried. If problem solvers could unite and push there with the proper solution elements, it appears the social side of the problem would be solved in short order, and civilization could at last enter the Age of Transition to Sustainability.

Political Powerplaces Are Everywhere

The first thing to understand about the Dueling Loops is they are a variation on an even more fundamental social structure: political powerplaces. These are ubiquitous, because our definition of a **political powerplace** is *any group where a leader's power depends on voluntary support rather than force*. The invention of democracy formalized a particular type of political powerplace: the elected politician. That is the powerplace this book is most concerned with, but other powerplaces can be just as important at times. One example is the deference and coverage support given to leading journalists, newscasters, and pundits. They are not elected. But their many supporters are voluntary and "vote" every time a viewer reads or watches a favorite source.

The basic political powerplace structure is shown. This also illustrates how feedback loops work. A **feedback loop** is a structural shape that causes output from one node to eventually influence input to that same node. The **Winning Supporters** loop has four nodes. Let's walk around the loop to see how it works.

Suppose a politician *wants* to win some supporters. She would offer proof of ability in the form of her voting record, the bills she has helped design or promote, and so on. If she had no time in office her proof might be her job experience, credentials, or well thought out positions on topics of concern to voters. The stronger her proof of ability, the more supporters she would get, as shown by the arrow connecting those two nodes. The more supporters she has, the greater her supporter's influence, such as in helping her get elected or mustering support for bills she supports. The more of that she has, the greater her leader power. The greater that is, the more chances she has to increase proof of ability, and the loop starts all over again.

As the loop goes round and around, it grows stronger and stronger. It's a **reinforcing loop** because a change in a node goes around the loop and causes a change in the same direction. As leader power goes up, that causes leader power to ultimately go up even more, due to the loop. The loop explains why incumbents have an inherent (and arguably unfair) advantage over those who are not in office.

The four nodes are not the whole story, however. A fifth node must also be considered, as shown in the expanded model. The enticements node represents the favors, promises, sweet talk, downplaying of their opponents, and so forth that politicians use to win supporters. These have nothing to do with proven ability to be a good public servant. Enticements work just as well as proof of ability to win supporters. Actually, as you will see later on, they work even better.

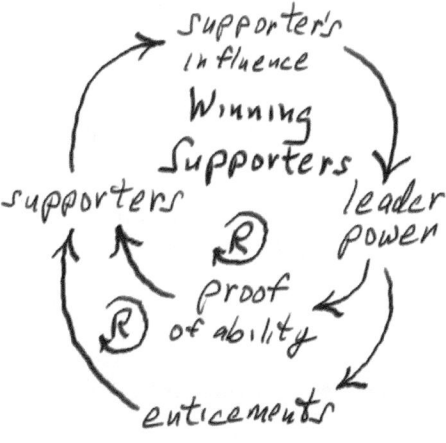

Now that we have added the enticements node, the structure explains the basic behavior of political powerplaces. They all have the same five nodes. Among other things, the structure explains why some leaders work so hard to maximize their power. It's because they want to maximize proof of ability and enticements so as to win the most supporters they can.

There is more detail we could add, but let's consider just the side effects of enticements on the rest of the system. The new model is shown below.

Enticements include favors to special interests, such as tax breaks, special exemptions, weak regulations, etc. The many laws allowing corporations to pollute more than they should are a prime example. After the *delay* represented by the double slash on the arrow going down from enticements, catering to special interests can cause negative side effects. An increase in the side effects will have an opposite effect on supporters and other supporters (those supporting other leaders). It will reduce the number of supporters, as health problems, food shortages, and conflict over scarce resources causes the population to fall. Inverse relationships are represented by a dashed arrow.

This creates the ***Losing Supporters*** loop. It's a **balancing loop** because it has an odd number of inverse relationships. This causes a change in the value of a node to eventually cause a change in that node in the *opposite* direction. If you follow the path of influence around, from supporters, to supporter's influence, to leader power, to enticements, to negative side effects and finally back to supporters, you will see that an *increase* in supporters eventually causes a *decrease* in supporters. By contrast, reinforcing loops have an ever number of inverse relationships, usually zero. Reinforcing and balancing loops are an important concept to grasp, because if they are well understood, problem solvers can use new or stronger loops to counter the destructive effects of feedback loops that are running out of control.

The power of a model stems from its explanatory and predictive ability. As simple as this one is, it already has enough power to explain why the environmental sustainability problem is running out of control. Once politicians get into power and start using the **Winning Supporters** loop, some will invariably notice that certain kinds of enticements gain them boatloads of supporters. Many of these enticements will cause negative side effects to the environment. But to the politician in power, that doesn't matter because of the *delay*. In the short run they will gain more supporters. So why not do whatever it takes to gain as many as possible?

Let's turn our attention to constructing the Dueling Loops model. As it grows, you will see it is no more that a variation on the basic structure of political powerplaces.

The Race to the Bottom

We hypothesize that that over time, cultural evolution has pared the many strategies available for gaining political support into just two main types: the use of truth (**virtue**) and the use of falsehood and favoritism (**corruption**). A third strategy, force, used to be an alternative. But the rise of democracy has mostly eliminated that.

Virtue and corruption are idealized endpoints on a spectrum. Strategies based on the truth seek to tell the public as close to the truth as realistically possible. Strategies near the other end of the spectrum do whatever it takes to get or stay elected.

Here's an example: A virtuous politician may gain supporters by stating, "I know we can't balance the budget any time soon, but I will form a panel of experts to determine what the best we can do is." Meanwhile, a corrupt politician is garnering supporters by saying, "Economics is easy. You just put a firm hand on the tiller and go where you want to go. I can balance the budget in four years, despite what the experts are saying. They are just pundits. Don't listen to them. A vote for me is a vote for a better future." The corrupt politician is also saying to numerous special interest groups, "Yes, I can do that for you. No problem." Guess who will usually win?

Winning in this manner is so much the norm that George Orwell wrote in *Politics and the English Language* that:

"Political language—and with variations this is true of all political parties, from Conservatives to Anarchists—is designed to make lies sound truthful and murder respectable, and to give appearance of solidity to pure wind." [22]

The use of corruption to gain supporters is the dominant loop in politics today. **Corruption** consists of falsehood and favoritism. Most politicians use rhetoric, half truths, glittering generalities, the sin of omission, biased framing, outright lies, and many other types of falsehood to make themselves look as appealing as possible to the greatest number of people.

Particularly when an election is drawing near, most politicians use the *ad hominem* (Latin for against the man) fallacy to attack and demonize their opponents. For example, the use of the Swift boat ads in the 2004 US presidential campaign to attack John Kerry's character were an *ad hominem* fallacy, because they had nothing to do with Kerry's political reasoning or positions. Other terms for the *ad hominem* fallacy are demagoguery, shooting the messenger, negative campaigning, smear tactics, and sliming your opponent. Finally, once in office nearly all politicians engage in acts of favoritism, also known as patronage.

Politicians are forced to use corruption to gain supporters, because if they do not they will lose out to those who do. This causes the **Race to the Bottom among Politicians** to appear, as shown on the next page.

To understand how the loop works, let's start at <u>false memes</u>. A **meme** is a mental belief that is transmitted (replicated) from one mind to another. Memes are a very useful abstraction for understanding human behavior because memes replicate, mutate, and follow the law of survival of the fittest, just as genes do. Rather than show falsehood and favoritism, the model is simplified. It shows only falsehood.

The more <u>false memes</u> transmitted, the greater the <u>degenerates infectivity rate</u>. The model treats arrival of

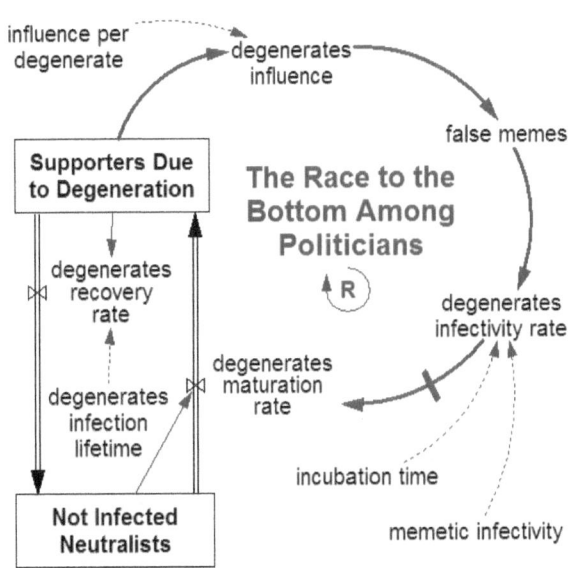

Structure of the Race to the Bottom

influence per degenerate

degenerates influence

false memes

Supporters Due to Degeneration

The Race to the Bottom Among Politicians

degenerates recovery rate

R

degenerates infectivity rate

degenerates maturation rate

degenerates infection lifetime

incubation time

Not Infected Neutralists

memetic infectivity

Figure 1. The loop grows in strength by using corruption in the form of highly appealing falsehood and favoritism. This increases the number of supporters of corrupt politicians, which increases their influence, which in turn increases their power to peddle still more falsehood and favoritism. Over time the loop can grow to tragically high levels.

a meme the same way the body treats the arrival of a virus: it causes infection. After the "mind virus" incubates for a period of time (a *delay*), the infection becomes so strong that maturation occurs. This increases the <u>degenerates maturation rate</u>, which causes supporters to move from the pool of <u>Not Infected Neutralists</u> to the pool of <u>Supporters Due to Degeneration</u> as they become committed to the false memes they are now infected with. <u>Supporters Due to Degeneration</u> times <u>influence per degenerate</u> equals <u>degenerates influence</u>. The more influence a degenerate politician has, the more <u>false memes</u> they can transmit, and the loop starts over again. As it goes around and around, each node increases in quantity, often to horrific levels. The loop stops growing when most supporters are committed.

The dynamic behavior of the loop is shown in the graph on the next page. The behavior is quite simple because the model has only a single main loop.

Corrupt politicians exploit the power of the race to the bottom by broadcasting as much falsehood and favoritism as possible to potential supporters. This is done with speeches, interviews, articles, books, jobs, lucrative contracts, special considerations in legislation, etc. The lies and favors are a cunning blend of whatever it takes to gain supporters. *The end justifies the means.* Note that the more influence a politician

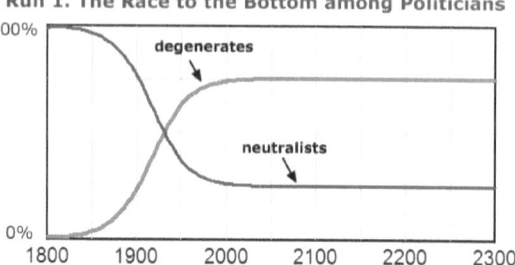

Run 1. The Race to the Bottom among Politicians

Figure 2. The simulation run starts with 1 degenerate and 99 neutralists. Over time the percentage of degenerates grows to 75% and stops. What keeps it from growing to 100% is the way degenerates can recover from their infection, after a <u>degenerates infection lifetime</u> of 20 years.

has, the more falsehood they can afford to broadcast, and the greater the amount of favoritism they can plausibly promise and deliver.

This is the loop that is driving politics to extremes of falsehood and favoritism in far too many areas of the world. This loop is the structural cause behind most of the corruption and bad decisions in government today.

The race to the bottom employs a dazzling array of deception strategies. These are usually combined, which increases their power. Here are the five main types of political deception:

1. False promise – A false promise is a promise that is made but never delivered, or never delivered fully. False promises are widely used to win and keep the support of various segments of the population, such as organized special interest groups, industries, and demographic groups like seniors or immigrants. False promises flow like wine during election season.

One of the largest false promises in recent history was the way Russian communism promised one thing but delivered another. It promised rule by the masses for the masses, but delivered a totalitarian state. To justify its continued existence and hide the broken promise, the communist system manufactured a steady stream of soothing lies and used harsh repressive techniques on those who did not swallow the lies.

Near the end of the collapse of Russian communism, Václav Havel, writing in 1978 in *Versuch, in der Wahrheit zu leben* (An Attempt to Live in Truth) pointed out the diabolical, self-destructive nature of the communist approach. It was the ultimate vicious cycle because:

"…it turned victims into accomplices: by threatening them and their descendents with disadvantages, it coerces the victims to participate. When Havel became President [of Czechoslovakia in 1989] he reminded his fellow citizens of their complicity arising from their coming to terms with life in lying. Consequently, he exhorted them… to vote for candidates who 'are used to telling the truth and do not wear a different shirt every week'." [23]

Civilization has a learning problem. *It does not seem to learn from its mistakes, even when they are pointed out.* It has not learned the lesson that false promises work so well to destroy lives *en masse* that their effectiveness must be eliminated somehow. This is nothing new, however. We have been warned before. For example, long ago in the 14th century Machiavelli explained why false promises are so rampant in *The Prince*, in the chapter on "How Princes Should Honor Their Word:"

"Everyone knows how praiseworthy it is for a prince to honor his word and to be straightforward rather than crafty in his dealings; nonetheless contemporary experience shows that princes who have achieved great things have been those who have given their word lightly, who have known how to trick men with their cunning, and who, in the end, have overcome those abiding by honest principles. …it follows that a prudent ruler cannot, and must not, honor his word when it places him at a disadvantage and when the reasons for which he made his promise no longer exist. … Everyone sees what you appear to be, few experience what you really are."

2. False enemy – Creating a false enemy works because it evokes the instinctual fight or flight syndrome. The brain simply cannot resist becoming aroused when confronted with a possible enemy.

The two main types of false enemies are *false internal opponents*, such as negative campaigning, the Salem witch trials, and McCarthyism, and *false external opponents*, such as communism and the second Iraq "war." While communism and Iraq were true problems, both were trumped up enormously to serve the role of a false enemy. False enemies are often scapegoats. A **scapegoat** is someone who is blamed for misfortune, usually as a way of distracting attention from the real causes or more important issues. Name-calling (such as tree huggers and tax-and-spend liberals) and ad hominem attacks are popular ways to create false enemies.

When it comes to creating false internal enemies, the winning strategy is to *attack early and attack often*. This becomes doubly successful when those attacked are politicians in the opposing party: (1) The fight or flight instinct is evoked, which clouds the judgment and causes people to want a strong milita- ristic leader to lead them out of harms way. The attacker proves his militaristic capability by the viciousness of his attack, causing those witnessing the attack to frequently swing their support to him. (2) Attacks cause the attacker's own supporters to fervently support him even more, because he has just pointed out why the opposition is so bad.

This form of deception works so well that attack politics has become *the* central strategy for many degenerate parties. Look around. Are there any po- litical parties whose most outstanding trait is they are essentially one gigantic, ruthless, insidiously effective attack machine?

3. Pushing the fear hot button – When a politician talks about almost everything in terms of terrorism, communism, crime, threats to "national secu- rity" or "our way of life," and so on, that politician is pushing the fear hot button. It's very easy to push. Just use a few of the right trigger words, throw in a dash of plausibility, and the subconsciousness is instinctively hoodwinked into a state of fear, or at least into wondering if there is something out there *to* fear. Whether or not an enemy actually *is* out there doesn't matter—what matters is that we think there *might* be one.

Fear clouds the judgment, making it all the harder to discern whether the enemy really exists. Because we cannot be sure, we play it safe and assume there is at least some risk. Since people are risk averse, the ploy works and we become believers. We have been influenced by statements of what *might* be lurking out there. Our fear hot button has been pushed and it worked.

How effective fear can be is echoed in this quote:

> "Fearful people are more dependent, more easily manipulated and con- trolled, more susceptible to deceptively simple, strong, tough measures and hard-line postures," [Gerbner] testified before a congressional subcommittee on communications in 1981. "They may accept and even welcome repression if it promises to relieve their insecurities. That is the deeper problem of violence-laden television." [24]

That was 1981. Today, little has changed. Al Gore, writing in *The Assault on Reason* in 2007, included an entire chapter on The Politics of Fear. It may as well have been called The Politics of Pushing the Fear Hot Button. Below are excerpts: (Italics added, except for the last use of "terrorism," which is italicized in the original. My comments are in brackets.)

> *"Fear is the most powerful enemy of reason.* Both fear and reason are essential to human survival, but the relationship between them is un-

balanced. Reason may sometimes dissipate fear, but fear frequently shuts down reason. As Edmond Burke wrote in England twenty years before the American Revolution, 'No passion so effectually robs the mind of all its powers of acting and reasoning as fear.'

"Our Founders had a healthy respect for the threat fear poses to reason. They knew that, under the right circumstances, fear can trigger the temptation to *surrender freedom to a demagogue promising strength and security in return*. [This is an example of a false promise.] They worried that when fear displaces reason, the result is often irrational hatred [which creates a false enemy] and division.

"Nations succeed or fail and define their essential character by the way they challenge the unknown and cope with fear. And much depends on the quality of their leadership. If leaders *exploit public fears to herd people in directions they might not otherwise choose*, [which is why they push the fear hot button] then fear itself can quickly become a self-perpetuating and freewheeling force that drains national will and weakens national character, *diverting attention from real threats....* [A wrong priority]

"It is well documented that *humans are especially fearful of threats that can be easily pictured or imagined*. For example, one study found that people are willing to spend significantly more for flight insurance that covers 'death by terrorism' that for flight insurance that covers 'death by any cause.' Now, logically, flight insurance for death by any cause would cover terrorism in addition to a number of other potential problems. But something about the buzzword *terrorism* creates a vivid impression that generates excessive fear." [Here terrorism has been used not only to push the fear hot button. It doubles as a way to create a false enemy.]

4. Wrong priority – Wrong priorities stem from hidden agendas. A **hidden agenda** is a plan or goal a politician must conceal from the public, due to an ulterior motive.

There are many ways a hidden agenda can come about. A politician may support a certain ideology, and so bends everything to support the goals of that ideology. He may have accepted donations and/or voter support from special interests, such as corporations, and therefore must promote their agenda. Perhaps he had to cut a deal.

A politician with a hidden agenda must make the wrong priorities seem like the right ones in order to achieve what's on the hidden agenda. How can he do this? For a corrupt politician such matters are child's play—manipulate the public through false promises, create a false enemy, push the fear hot button hard and often, repeat the same lie over and over until it becomes "the truth," and so forth.

The low priority that environmental sustainability receives from most governments today is rapidly becoming *the* textbook example of how devastating wrong priorities can be.

5. Secrecy – The fifth main type of deception is actually a way to make the other four types ten times as easy to achieve. **Secrecy** is hiding or withholding the truth. It's a powerful form of deception because it creates a false impression without actually having to openly lie about anything. Secrecy makes it impossible to tell if a politician is lying, because key premises cannot be tested. One type of lie is *the sin of omission.*

Secrecy is so important to the success of the other four types of deception, that without it they would crumble into ineffective mumblings. But with secrecy they work most of the time, because there is no way for the population to tell if a politician is telling the truth or not. When you see a politician, administration, or party using much more secrecy than normal, and there is no reasonable justification, you can be certain that its purpose is deception.

The use of secrecy by politicians for nefarious ends is so pervasive that numerous books have appeared on the subject. Recent examples are:

- Secrecy: Political Censorship in Australia, Spigelman, 1972.
- Executive Privilege: The Dilemma of Secrecy and Democratic Accountability, Rozell, 1994.
- The Culture of Secrecy: Britain, 1832–1998, Vincent, 1998.
- *Secrecy: The American Experience*, Senator Daniel Patrick Moynihan, 1999.
- Secrecy Wars: National Security, Privacy, and the Public's Right to Know, Melanson, 2002.
- Command of Office: How War, Secrecy, and Deception Transformed the Presidency from Theodore Roosevelt to George W. Bush, Graubard, 2004.

While it examines only one country, the Moynihan book penetrates to the root of why excessive secrecy is inherently bad: it allows governments to pursue ends that are not in the best interests of the people. As Moynihan sees it, the greater threat is not the one you might expect: keeping secrets from the public. It is the keeping of secrets between units of government, which occurs when:

The Sustainability Problem as a Running Example

Of all the problems facing progressives, currently the top long term problem is the global environmental sustainability problem. If we can solve this one then we can solve any of them, because they are all complex social system problems, they all appear to be the result of exploitation of the race to the bottom, and this one is representative and probably the most difficult problem of them all.

For these reasons this book uses the sustainability problem as an educational running example of how any of the problems that make up the paradox can be solved.

"Departments and agencies hoard information, and the government becomes a kind of market. Secrets become organizational assets, never to be shared save in exchange for another organization's assets. Sometimes the exchange is in kind: I exchange my secret for your secret. Sometimes the exchange resembles barter: I trade my willingness to share certain secrets for your help in accomplishing my purposes. But whatever the coinage, the system costs can be enormous. In the void created by absent or withheld information, decisions are either made poorly or not made at all." [25]

Moynihan shows in convincing detail how this pattern of dysfunctional secrecy led to the Cold War, when it could have been avoided. How it led to the excesses of McCarthyism and the Bay of Pigs fiasco, when both could also have been avoided. And how, if it continues, a "culture of secrecy" will most assuredly lead to more unnecessary wars and other tragedies.

Clever Rationalizations – The five main types of political deception won't work at all unless they can be implemented. The most common implementation technique is to rationalize why a false promise is really true, why a false enemy is real, why there is a bogyman to fear, why the wrong priority is really the right priority, and why secrecy is necessary when it's really not.

A **rationalization** is a falsehood supporting a pre-conceived conclusion or goal. The best rationalizations are the result of extensive testing and competition with other rationalizations, such as by testing on focus groups or small markets. All rationalizations employ well known fallacies to trick the receiver into believing a statement is true, when in fact it is false.

For example, the widely circulated argument that the Kyoto Protocol would not solve the climate change problem, and therefore is not worth supporting, is a clever rationalization. Of course it won't solve it, because the first round of greenhouse gas emission reductions (averaging 5.2% below 1990 levels) are only a first step. Another popular rationale is that mandatory emission limits would harm the US economy. It is true that GDP will probably fall as lower amounts of fossil fuels, cars, trucks, and so on are consumed. But the long term harm will be much greater if nothing was done. Yet another rationale is why should the US support the treaty if China and India are exempt? The false answer is the US should not. But the true answer is the less developed countries will be included in later phases of the treaty. It makes little sense to include them in the early phases, because they are not a major source of emissions per capita now (except for exceptions like China), nor have they been a major source in the past.

Other Types – There are many more ways to implement the five main types of deception, such as biased framing, spin, false grassroots organizations, biased "public relations," false advertising, false news stories, the fallacy of "balanced news," casting doubt on the severity or urgency of a problem, etc.

The right steady drumbeat of false promises, false enemies, pushing the fear hot button, wrong priorities, secrecy, and clever rationalizations creates the ultimate political weapon: lies that work on entire nations. This is why history has given us these gems of dark wisdom:

> *"Next the statesmen will invent cheap lies, putting the blame upon the nation that is attacked, and every man will be glad of those conscience-soothing falsities, and will diligently study them, and refuse to examine any refutations of them; and thus he will by and by convince himself that the war is just, and will thank God for the better sleep he enjoys after this process of grotesque self-deception."* – Mark Twain, *The Mysterious Stranger*, 1910.

> *"The whole aim of practical politics is to keep the populace alarmed (and hence clamorous to be led to safety) by menacing it with an endless series of hobgoblins, all of them imaginary."* – H. L. Mencken, *In Defense of Women*, 1917.

> *"A lie repeated often enough becomes the truth."* – Vladimir Lenin.

"It does not matter how many lies we tell, because once we have won, no one will be able to do anything about it." – Statement by Dr. Joseph Goebbels to Adolf Hitler, early 1930s, from *The Rise and Fall of the Third Reich*, by William L Shirer.

More modern history has given us this one:

"The Greatest Story Ever Sold: The Decline and Fall of Truth from 9/11 to Katrina – The title of a 2006 book by Frank Rich. A review in the New York Times gives us a deeper look at Rich's message: [26]

"The truly cynical political operator, whether Republican or Democrat, could read this book as a manual for how to use deception, misinformation and propaganda to emasculate your enemies, subdue the news media and befuddle the public, and not as the call to arms for truth that Mr. Rich seeks to provide."

It sounds like Machiavelli is alive and well, and working as a consultant to any government who agrees that *the ends justify the means*. Notice Rich's intuitive realization that the "Fall of Truth" is the cause of the corruption problem currently haunting America, and a "call to arms for the truth" is the cure. This leads to what Henry David Thoreau wrote in *A Week on the Concord and Merrimack Rivers,* in 1849:

"It takes two to speak the truth—one to speak, and another to hear."

Which in turn leads to our own observation:

"It takes two to speak the lie—one to speak, and one to be deceived."

Below is a summary of the five main types of political deception from the video series on the Dueling Loops at Thwink.org. Notice the many ways to implement the five types. [27]

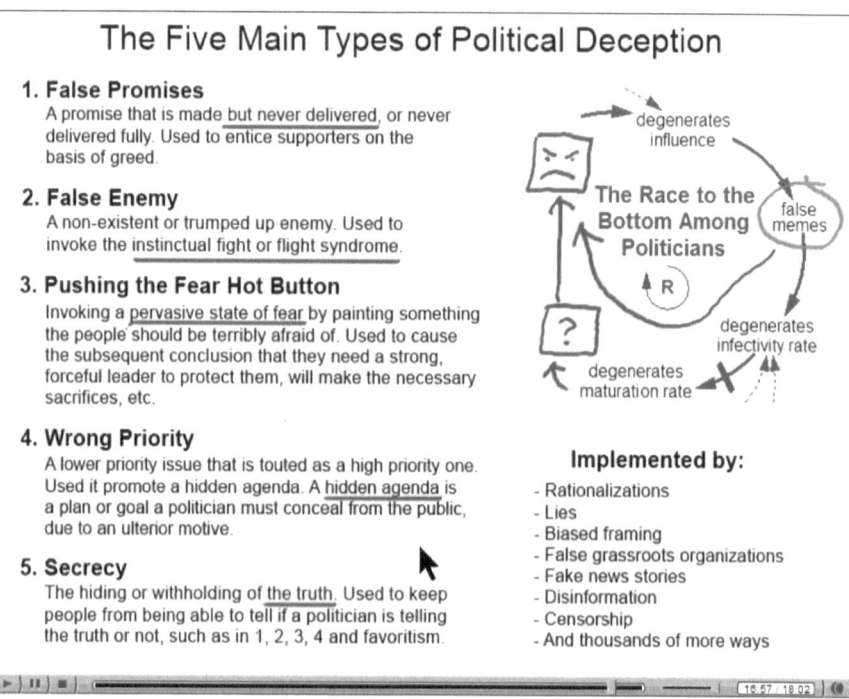

The Five Main Types of Political Deception

1. False Promises
A promise that is made but never delivered, or never delivered fully. Used to entice supporters on the basis of greed.

2. False Enemy
A non-existent or trumped up enemy. Used to invoke the instinctual fight or flight syndrome.

3. Pushing the Fear Hot Button
Invoking a pervasive state of fear by painting something the people should be terribly afraid of. Used to cause the subsequent conclusion that they need a strong, forceful leader to protect them, will make the necessary sacrifices, etc.

4. Wrong Priority
A lower priority issue that is touted as a high priority one. Used it promote a hidden agenda. A hidden agenda is a plan or goal a politician must conceal from the public, due to an ulterior motive.

5. Secrecy
The hiding or withholding of the truth. Used to keep people from being able to tell if a politician is telling the truth or not, such as in 1, 2, 3, 4 and favoritism.

Implemented by:
- Rationalizations
- Lies
- Biased framing
- False grassroots organizations
- Fake news stories
- Disinformation
- Censorship
- And thousands of more ways

Opposing the race to the bottom is the race to the top. The two loops are joined together as shown on the next page. Because each loop competes for the same Not Infected Neutralists, they are "Dueling Loops."

In the **race to the top** virtuous politicians compete for supporters on the basis of the truth (on the model this is called true memes). No favoritism is used, because those who tell the truth treat everyone equitably. Virtuous politicians can help improve things so that society benefits as a whole, but they cannot promise or give anyone more than their fair share.

The race to the top works in a similar manner to the race to the bottom because the two loops are entirely symmetrical, with one crucial difference: in the race to the top, the size of the truth cannot be inflated. Corrupt politicians can use false meme size to inflate the appeal of what they offer their supporters. But virtuous politicians cannot use falsehood to promise more than they can honestly expect to deliver. Nor can they use favoritism to inflate expectations of how well they can help particular supporters.

The Basic Structure of the Dueling Loops

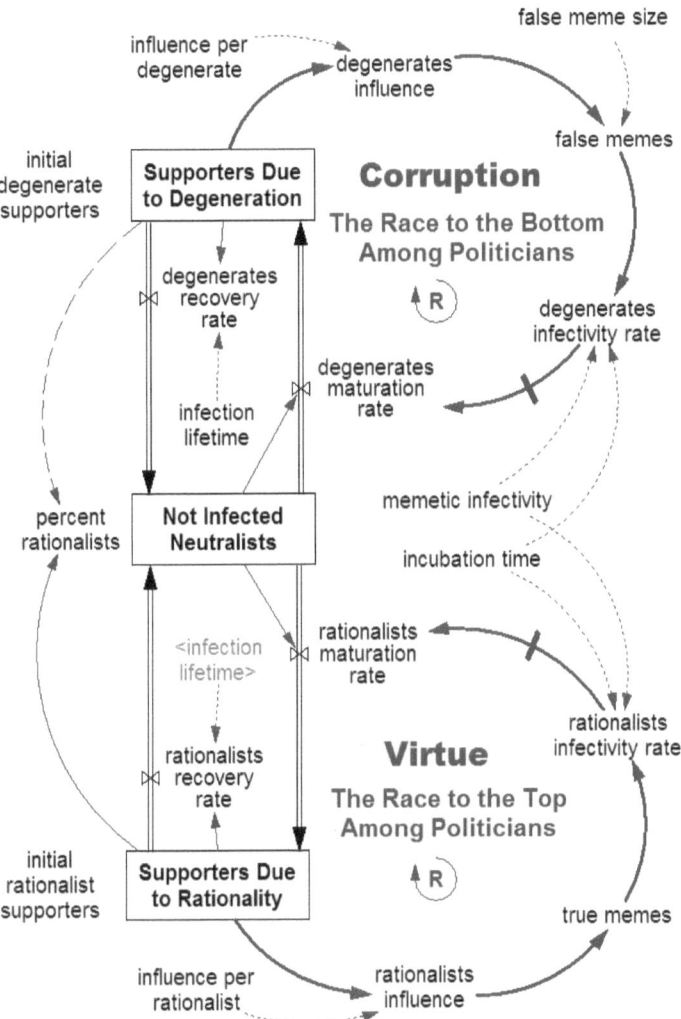

Figure 3. This is the basic structure of the dueling loops of the political powerplace. There are many variations. This structure, combined with agent selfishness, is the fundamental cause behind the behavior of all political systems, both ancient and modern. In particular this structure explains why corruption is what dominates politics, no matter how hard society tries to stamp it out. But once the structure is deeply understood it becomes possible to arrive at a way to eliminate corruption indefinitely. This is required to achieve sustainability of any kind, because **sustainability** is defined as the ability to continue a defined behavior indefinitely.

Why exactly do virtuous politicians feel they cannot tell lies? The goal of virtuous politicians is to optimize the common good for all, which includes those who will follow us. The common good includes *the rule of telling the truth*, because the more you can assume a person is telling the truth, the more effectively you can cooperate. Effective cooperation is the foundation upon which all social contract societies are built. Because virtuous politicians feel compelled to tell the truth, they avoid lying. They know that if they start telling lies their society will begin to crumble. Eventually it will degrade to life in mankind's natural state (before that of a central government based on cooperation) where, as Thomas Hobbes put it, "the life of man" is "nasty, brutish, and short."

But corrupt politicians feel no such constraint. Their goal is the uncommon good, that is, the good of special interests. Instead of the rule of telling the truth, corrupt politicians follow *the rule of expediency:* do whatever it takes to maximize the good of the special interests supporting you. The end justifies the means. If a situation is best exploited by telling the truth, tell it. If it's best exploited by a combination of truth and lies, then do that. This makes it impossible to trust corrupt politicians. But that doesn't matter, because if their deception is successful the public has no idea they are being exploited.

By examining how the basic dueling loops model behaves in a series of simulation runs, we can better understand why the political powerplace works the way it does. The table below lists the first six simulation runs we will examine. The first two variables are the **changeable variables**. By varying the changeable variables from run to run, we can try different scenarios. Each scenario is a logical experiment. The third variable is a **result variable**. It is the outcome of a simulation run, after equilibrium is reached.

Basic Dueling Loops Model Variables	Simulation Runs					Table 1
	1	2	3	4	5	6
Initial rationalist supporters	0	1	5	1	1	1
False meme size	1	1	1	1.1	1.3	2
Percent rationalists	0%	50%	83%	20%	5%	0%

Run 1 – This was presented earlier in figure 2. By setting <u>initial rationalist supporters</u> to zero and <u>false meme size</u> to 1, we get the equivalent of the race to the bottom loop and graph that was presented earlier. <u>Initial degenerate supporters</u> equals 1 in all six runs.

Run 2 – In run 2 the number of <u>initial rationalist supporters</u> is increased to 1. Now both loops have the same number of initial supporters. Because neither loop has an advantage over the other loop, the result is both

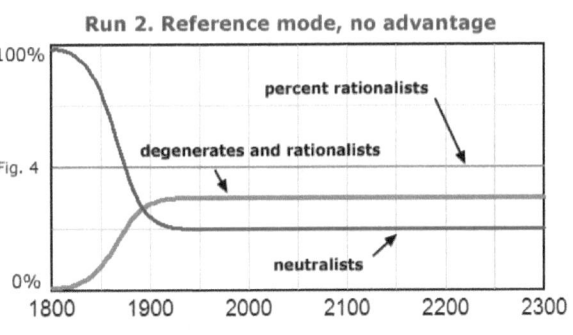

Run 2. Reference mode, no advantage

loops behave the same. Each attracts the same percentage of supporters, as shown below:

Because this run exhibits the most basic behavior of the dueling loops, without the whistles and bells of giving one side an advantage, it's our reference mode. A **reference mode** is what modelers use to compare all other runs to, because it is the most fundamental run or represents the current system. Notice how in this run the percentage of degenerates and rationalists are always the same, so the degenerates' curve covers the rationalists' curve. Both curves will be seen in later runs. <u>Percent rationalists</u> is the number of rationalists divided by degenerates plus rationalists. Naturally the higher this percentage is the better. In this run <u>percent rationalists</u> is always 50%.

Run 3 – In this run we increase initial rationalists to 5. This shows what happens if we give one side a head start on their number of supporters. Because we have not changed false meme size, neither size has an inher-

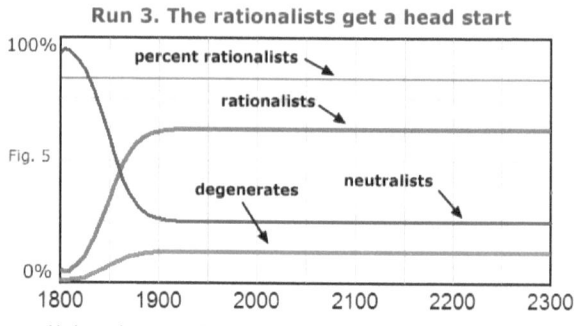

Run 3. The rationalists get a head start

ent advantage. But even a small head start, if all else is equal, can quickly become a large advantage, as the results show.

Run 4 – Now things get interesting. The number of initial rationalist supporters is set back to 1 and false meme size is increased from 1 to 1.1. This is only a tiny bit bigger, by 10%. It would seem that itsy bitsy lies and favors wouldn't make much difference, but no—they make a huge difference over a long period of time. As the run 4 graph below shows, the rationalists get wiped out. After 500 years they are down to about 20%. After 5,000 years (not shown) they are down to 0.345879 persons, which in the real world would be zero.

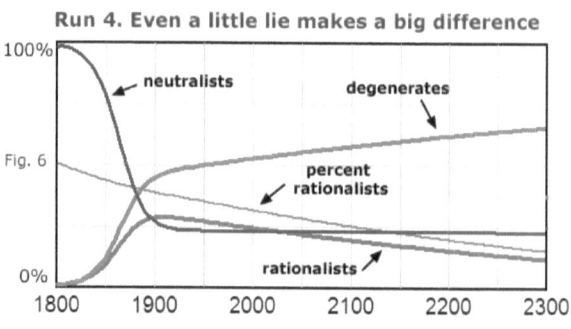

Run 4. Even a little lie makes a big difference

But notice how slowly the lines for degenerates and rationalists diverged for the first 50 years. What might happen if the degenerates decided to tell bigger lies and give out bigger favors?

Run 5 – If false meme size is increased from 1.1 to 1.3, system behavior changes dramatically. It only takes about 30 years for the degenerates to pull away from the rationalists. Now the degenerate and rationalist lines flatten out after only 500 years, instead of the 5,000 years it took in run

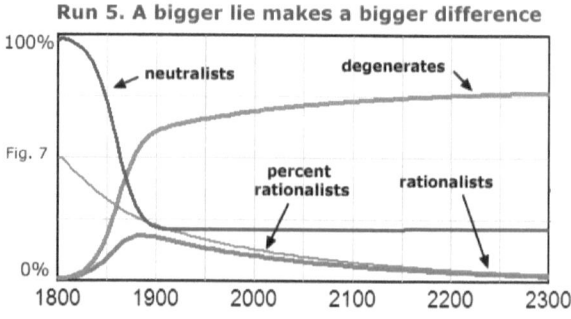

Run 5. A bigger lie makes a bigger difference

4. The end result is the same. The lesson is that the bigger the lie, the faster a corrupt politician can take over a political system. I wonder if that explains anything we might be seeing in politics today, such as in the United States?

Run 6 - Finally we see what happens if a corrupt politician decides to tell real whoppers. False meme size has increased to 2. In other words, every false promise, every false enemy, and so on is now twice as big as they really are.

The results are no surprise. Now the system responds so fast the rationalists never even make much of an impact on politics. They are smothered so fast by such big lies that the graph line for rationalists is starting to look like a

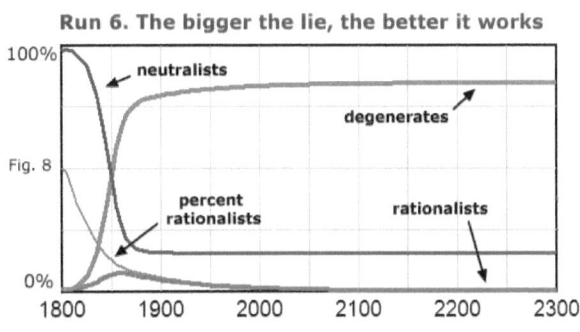

Fig. 8

pancake. Now, after only 500 years, there are 0% rationalists left in the system. They have been exterminated.

There is a limit to how big a lie can grow before it starts to make detection easy. In Figure 9 we will add the <u>effect of size of lie on detection</u> variable to the model, which will impose diminishing returns on the size of a lie.

This then is the basic structure of the dueling loops of the political power-place. The two loops are locked in a perpetual duel for the same <u>Not Infected Neutralists</u>. In addition, each politician has his or her own loop, and battles against other politicians for the same supporters. It is these many loops and the basic dueling loops structure that forms the basic structure of the modern political powerplace. The outstanding feature of this structure is:

The Inherent Advantage of the Race to the Bottom

Because the size of falsehood and favoritism can be inflated, and the truth cannot, the race to the bottom has an inherent structural advantage over the race to the top. This advantage remains hidden from all but the most analytical eye.

A politician can tell a bigger lie, like budget deficits don't matter. But they cannot tell a bigger truth, such as I can balance the budget twice as well as my opponent, because once a budget is balanced, it cannot be balanced any better. [28] From a mathematical perspective, the size (and hence the appeal) of a falsehood can be inflated by saying that $2 + 2 = 5$, or 7, or even 27, but the size of the truth can never be inflated by saying anything more than $2 + 2 = 4$.

Because the size of falsehood and favoritism can be inflated and the truth cannot, corrupt politicians can attract more supporters for the same amount of effort. A corrupt politician can promise more, evoke false enemies more, push the fear hot bottom more, pursue wrong priorities more, and use more favoritism than a virtuous politician can. *The result is the race to the bottom is normally the dominant loop.* Thus the reason that "Power corrupts, and absolute power corrupts absolutely" is not so much that power itself corrupts, but that the surest means to power requires corruption. [29]

Due to lack of an in-depth analysis of the fundamental causes of the social side of the problem, problem solvers have long been intuitively attracted to the low leverage point of pushing on **more of the truth**. On the model this point is the true memes node. The truth is discovered by research on technical ways to live more sustainably, such as population control, alternatives to fossil fuels, and reduce, reuse, and recycle. The truth is then spread by scientific reports, popular articles, environmental magazines, lobbying, pilot projects, lawsuits to enforce the legal truth, demonstrations to shock the public into seeing the real truth, and so on. This works on problems with low change resistance, such as local pollution problems and conservation parks. But it fails on those with high change resistance, like climate change, because environmentalists simply do not have the force (wealth, numbers, and influence) necessary to make pushing on this point a viable solution.

Because of its overwhelming advantage, the race to the bottom is the surest way for a politician to *rise to* power, to *increase* his power, and to *stay* in power. But this is a Faustian bargain, because once a politician begins to use corruption to win, he joins an anything goes, the-end-justifies-the-means race to the bottom against other corrupt politicians. He can only run faster and keep winning the race by increasing his corruption. This is why the race to the bottom almost invariably runs to excess, and causes its own demise and collapse.

This collapse ends a cycle as old as the first two politicians. A cycle ends when corruption becomes so extreme and obvious that the people rise up, throw the bums out, and become much harder to deceive for awhile. But as good times return, people become lax, and another cycle begins. These cycles never end, because presently there is no mechanism in the human system to keep ability to detect deception permanently high.

The dueling loops structure offers a clear explanation of why environmentalists are facing such a hostile political climate. This strong opposition occurs because a dominant race to the bottom causes corrupt politicians to work mostly for the selfish good of degenerate supporters, instead of working for the common good of the people. In other words:

The Race to the Bottom
Is Easily Exploited by Special Interests

Exploitation is the use of others to increase your own competitive advantage, at the cost of theirs. Because this is so obviously self-destructive to those being exploited, deception is required to pull it off. (We are considering only voluntary exploitation, and not cases like slavery.)

The race to the bottom provides the perfect mechanism for political exploitation, via election support of some type in return for favors. A little of this goes a long way, because each politician has his or her own loop. There are also hierarchies of loops, since a politician's supporters can be other politicians. At the top of each hierarchy is the top politician, such as a president, political strategist, or party. Whoever is at the top has tremendous leverage. *Thus the race to the bottom greatly amplifies the power of the exploiter.*

In stark contrast, the race to the top cannot be exploited. Unseemly rewards cannot flow to a truth telling politician without everyone knowing about it, because part of telling the truth is keeping no secrets and not committing the "sin of omission," a type of lie. Nor can the race to the top be exploited by supporters or outsiders with bribes or favoritism, because truth telling politicians would say no and if necessary report them. If they didn't, they would lose supporters because they would be committing falsehood.

Basically the race to the top is not exploitable because exploitation requires unjustified support, which is what the race to the bottom thrives on. But in the race to the top, all support is justified because it is based on the truth and the equitable distribution of the benefits of social cooperation.

The incentive to exploit occurs when a special interest group has interests that conflict with those of society as a whole. Common examples are religious fundamentalists, the rich, the military, and large corporations. The latter two make up the infamous military industrial complex.

A corrupt politician, by accepting donations (legal bribes) and votes in return for favoritism, becomes beholden to the special interest groups involved. If a special interest is powerful enough, it can control and exploit a political system by clever use of the race to the bottom. This is exactly what is happening today. The global political system is by and large being exploited by:

Chapter 5

The New Dominant Life Form

L ET'S DEFINE A **LIFE FORM** AS ANY INDEPENDENT AGENT THAT FOLLOWS the three fundamental requirements of evolution. These requirements are replication, mutation, and survival of the fittest.

Here's a question: What life form has the ability to replicate instantly with almost no expenditure of energy, can mutate during replication or at any time thereafter, and, when it has failed in the battle of survival of the fittest, sells little pieces of itself to its competitors in order to minimize its own pain of death? These are fantastic powers no human could hope to have. But what if we go further, and ask what life form has the miraculous power of being in many places at the same time, has an infinite life span, and can cleave off chunks of itself and have them instantly come alive? That would make it a formidable competitor indeed, one that could run rings around any other plant or animal. Darwin would be astounded.

But there's more. What life form totally dominates mankind, by controlling most jobs in developed countries, by determining the path of nearly all of new technology, products, and services, by controlling elections and political decisions more than any other life form, and by defining the very evolution of culture to its advantage through demand advertising, ownership of the media, and new product design? If that is not enough, what life form controls the billions of boxes in our homes that provide us with most of our "news," and most of our new knowledge once we have finished school, while at the same time subconsciously indoctrinating us to be high volume, complacent consumers? To top it off, what life form is spreading exponentially from industrialized countries to the rest of the world, and will soon dominate them all? The answer is obvious: It is the modern corporation, which is the New Dominant Life Form.

Thus the dominant life form on Earth is no longer Homo sapiens. Instead, it is the modern corporation and its allies. [30]

This is the real force progressives and environmentalists are battling. The second Bush administration, as well as others before it and around the world who oppose sustainability, are mere proxies for the real opponent: the modern corporation and its allies. Its allies include many of the rich, the military, politicians, and special interest groups, such as the religious right.

Please note this is not an indictment of all corporations and their managers. Most are doing the best they can, and are basically good. Each agent, from its own perspective, is behaving rationally. It is the life form as a whole that

The World's 100 Largest Economies

Corporate revenues versus country GDP for 2000 in millions of US$

#			#		
1	United States	$9,882,842	51	Iran	$98,991
2	Japan	$4,677,099	52	Egypt	$98,333
3	Germany	$1,870,136	53	Ireland	$94,388
4	United Kingdom	$1,413,432	54	Axa	$92,781
5	France	$1,286,252	55	Singapore	$92,252
6	China	$1,079,954	56	Sumitomo	$98,168
7	Italy	$1,068,516	57	Malaysia	$89,321
8	Canada	$689,550	58	IBM	$88,396
9	Brazil	$587,553	59	Marubini	$85,351
10	Mexico	$574,512	60	Colombia	$82,849
11	Spain	$555,004	61	Volkswagen	$78,851
12	India	$479,404	62	Hitachi	$76,126
13	South Korea	$457,219	63	Philippines	$75,186
14	Australia	$394,023	64	Siemens	$74,858
15	Netherlands	$364,948	65	ING Group	$71,195
16	Argentina	$285,473	66	Allianz	$71,022
17	Russian Federation	$251,092	67	Chile	$70,710
18	Switzerland	$240,323	68	Matsushita	$69,475
19	Belgium	$231,016	69	E.ON Energy	$68,432
20	Sweden	$227,369	70	Nippon Life Insurance	$68,054
21	ExxonMobil	$210,392	71	Deutsche Bank	$67,133
22	Turkey	$199,902	72	Sony	$66,158
23	Wal-Mart	$193,295	73	AT&T	$65,981
24	Austria	$190,957	74	Verizon	$64,707
25	General Motors	$184,632	75	U. S. Postal Service	$64,540
26	Ford	$180,598	76	Philip Morris	$63,276
27	Hong Kong	$163,261	77	Pakistan	$61,673
28	Denmark	$160,780	78	CGNU	$61,498
29	Poland	$158,839	79	J. P. Morgan & Chase	$60,065
30	Indonesia	$153,255	80	Carrefour	$59,887
31	DaimlerChrysler	$150,069	81	Credit Suisse	$59,315
32	Norway	$149,349	82	Nissho Iwai	$58,557
33	Royal Dutch/Shell	$149,156	83	Honda	$58,461
34	BP	$148,062	84	Bank of America	$57,747
35	General Electric	$129,853	85	BNP Paribas	$57,611
36	Mitsubishi	$126,579	86	Nissan	$5,077
37	South Africa	$125,887	87	Peru	$53,882
38	Thailand	$121,927	88	Toshiba	$53,826
39	Toyota	$121,416	89	Algeria	$53,817
40	Venezuela	$120,484	90	PDVSA	$53,680
41	Finland	$119,823	91	Assicuraz. Generali	$53,333
42	Mitsui	$118,013	92	Fiat	$53,190
43	Greece	$111,955	93	Mizuho	$52,068
44	CitiGroup	$111,826	94	SBC Communications	$51,476
45	Israel	$110,332	95	Boeing	$51,321
46	Itochu	$109,765	96	Texaco	$51,130
47	Total FINA Elf	$105,869	97	New Zealand	$49,943
48	Portugal	$103,871	98	Fujitsu	$49,603
49	NTT	$103,234	99	Czech Republic	$49,510
50	Enron	$100,789	100	Duke Energy	$49,318

In terms of corporate revenues vs national gross domestic product (GDP), of the 100 largest economies in the world in the year 2000, 53 were corporations. [31]

has the emergent property of behaving unsustainably. [32]

Avoiding the Fundamental Attribution Error

Because the New Dominant Life Form is behaving so unsustainably and appears to be the major source of change resistance to solving the sustainability problem, there is a tendency for environmentalists to demonize corporations and their managers, and call them the cause of the problem. This is a serious error, because it is not they who are at fault. It is the overall structure of the system that is causing them, on the average, to behave the way they do. *Thus it is the system that is at fault, not corporations.*

This error is so common it has become known among behaviorists as the fundamental attribution error. So that you can become a better structural thinker, here's what John Sterman, writing in *Business Dynamics: Systems Thinking and Modeling for a Complex World*, has to say on this topic: (Italics and bolding added)

> "A fundamental principle of system dynamics states that *the structure of the system gives rise to its behavior.* However, people have a strong tendency to attribute the behavior of others to dispositional rather than situational factors, that is, to character and especially to character flaws rather than the system in which these people are acting. The tendency to blame the person rather than the system is so strong psychologists call it the '**fundamental attribution error.**'
>
> "In complex systems different people placed in the same structure tend to behave in similar ways. When we attribute behavior to personality we lose sight of how the structure of the system shaped [their] choices. The attribution of behavior to individuals and special circumstances diverts our attention from the *high leverage points* where redesigning the system or governing policy can have significant, sustained, beneficial effects on performance. When we attribute behavior to people rather than system structure the focus of management becomes *scapegoating and blame*, rather than the design of organizations in which ordinary people can achieve extraordinary results." [33]

Note the last sentence. Committing the fundamental attribution error causes us to lose sight of where we should be directing our efforts, and become bogged down in "scapegoating and blame" instead. Better is to keep our eyes on the prize, which is to use our problem solving skills to redesign the system so that it works as intended. We must remember that "Take a good person and put them in a bad system, and the system wins every time." [34]

If activists can avoid the knee jerk reaction of the fundamental attribution error they will see that corporations are not the cause of the problem after all. That particular social agent is only a superficial cause. To find the true root cause we must go much deeper.

We need a name for this superficial cause. The term **New Dominant Life Form** means just that and no more. It was not coined to serve as a derogatory label, as some readers of drafts of this book have reacted. It is judgmentally neutral, just as the term **Previous Dominant Life Form** for *Homo sapiens* would be. [35]

How Dominant?

Global dominance by the corporate life form has been achieved by a centuries long series of incremental steps.

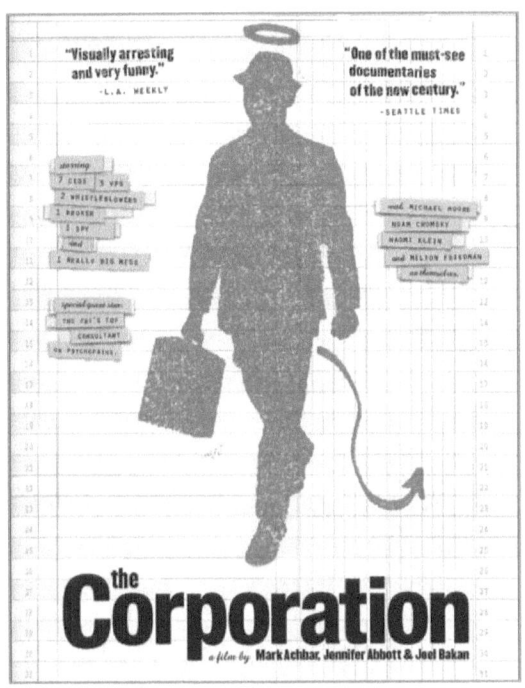

The above film falls into the fundamental attribution error trap by blaming *The Corporation* for many of society's problems, rather than the structure of the system. The DVD cover and much of the film's content "demonize the enemy" and are thus an appeal to emotion rather than reason. Because appeal to emotion is a fallacious argument, this film contributes to The Race to the Bottom among Politicians. It therefore does more harm than good.

Taken alone, none have been so objectionable as to be blocked. Thus is was that the New Dominant Life Form crept up on its rival, *Homo sapiens*, silently and undetected. Then, on January 1, 1995 it pounced.

That was the day the **World Trade Organization** (WTO) was born. A more accurate name would be the World Government by Corporations, because its primary purpose is to maximize the primary energy input (money via sales) of the New Dominant Life Form. This is done by maximizing international trade, at the expense of all other system behavior and life forms. David Korten, writing in *When Corporations Rule the World*, points out that:

"The key provision in the 2,000 page agreement creating the WTO is buried in paragraph 4 of Article XVI: 'Each member [nation] shall

ensure the conformity of its laws, regulations, and administrative pro-
cedures with its obligations as provided in the annexed agreements.'

"The 'annexed agreements' include all the substantive multilateral
agreements relating to trade in goods and services and intellectual
property rights. This provision allows a WTO member country to challenge any law of another member country that it believes deprives it of
benefits it expected to receive from the new trade rules. This includes
virtually any law that requires imported goods to meet local or national health, safety, labor, or environmental standards that exceed
WTO accepted international standards." [36]

Oh my gosh is all I could say when I read that. What have we done? But it
gets worse, because the WTO has absolute powers of judgment and enforcement, to which there is no appeal. Sharon Beder, in *Suiting Themselves: How
Corporations Drive the Global Agenda*, 2006, describes these powers:

"Today, the WTO has greater powers than any other international institution, including powers to punish non-complying nations that are
not even available to the United Nations. Over 130 nations are now
members of the WTO. It has become a form of global government in
its own right with judicial, legislative, and executive powers.

"The WTO has come to rival the International Monetary Fund as
the most powerful, secretive, and anti-democratic international body
on Earth. It is rapidly assuming the mantle of a bona fide global government for the 'free trade era,' and it actively seeks to broaden its
powers and reach.

"The WTO is able to enforce its rules through its dispute settle-
ment mechanism. If a country complains that another is not abiding by
WTO rules, the case is heard by panels of unelected lawyers and officials…, behind closed doors with no public scrutiny. These panels are
able to find countries guilty of breaking the rules and to impose economic sanctions as punishments.

"Such rulings can declare legislation put in place by democrati-
cally elected governments as illegal. The WTO has fairly extensive
powers to discipline nation states—as well as local, state, and regional
governments—for regulations and controls that are claimed to inter-
fere with trade. WTO rules also take precedence over other international agreements, including labor and environmental agreements…" [37]

With dominance like this, is it any surprise that the sustainability problem
has played out the way it has?

The Inevitable Consequence of Mutually Exclusive Goals

The goal of an agent determines its behavior. The goal of most for-profit corporations is to maximize the net present value of profits. The goal of most people, once they have gotten past the survival and security stage, is to maximize quality of life for themselves and their descendants.

These goals are mutually exclusive. As a result, as things get better for the New Dominant Life Form they get worse for the Previous Dominant Life Form: *Homo sapiens*. For example, as Gross World Product continues to rise, sales and profits soar to unprecedented heights. However, so does pollution and natural resource depletion. While their effects are delayed, it is only a matter of time before the quality of life for *Homo sapiens* begins to fall.

Previously corporations were artificially created entities designed to serve their masters: people. But now the relationship has been reversed. It is the modern corporation who is now the master, and people are its servants. But because most people cannot find another master to work for, they have no choice but to work for this one.

When two life forms compete for dominance of the same ecological niche, the one with the most competitive advantage wins. Please page 145 for a table comparing the competitive advantage of the corporate life form to *Homo sapiens*. Note the lopsided advantage of corporations.

It is a paradox why *Homo sapiens* would create an entity that is more powerful that itself and has a mutually exclusive goal. Such a creation is guaranteed to cause its creator great harm, if not eventual extinction. But it is really not a paradox at all—it is an experiment gone awry. So awry, in fact, that it is time to end the experiment by redesigning that creation....

The Deeper Question

It's easy to jump to the conclusion that all we have to do to resolve the problematic behavior of the New Dominant Life Form is to redesign the modern corporation so that its goals are no longer in conflict with that of *Homo sapiens*. But this does not strike at the root. Furthermore, intuitive tampering with this particular agent's design could too easily backfire, and cause huge unintended consequences.

The deeper question to ask is WHY are key agents like corporations misdesigned? Later in this book the Niche Succession model identifies a root cause as low quality of political decision making. If this can be raised to a high level, then the system will now seek to reengineer corporations so that they serve their human masters rather than themselves. From a structural thinking viewpoint, this must be done in conjunction with other changes to the system, which will be considerable. One example, discussed later in this book, is raising general ability to detect political deception. Another is it may be necessary to design new public servants to handle crucial new system roles, such as environmental property management. (This was later renamed to be common property rights.)

The Root Cause of Change Resistance

We now have enough pieces of the puzzle to draw an important conclusion: The dueling loops, their cyclic nature, the inherent advantage of the race to the bottom, the presence of the New Dominant Life Form, and its successful exploitation of the race to the bottom are the **structural root cause** of most of the stiff, prolonged resistance to adopting a solution to the environmental sustainability problem. Civilization is presently stuck in the dominant race to the bottom part of the cycle. *Our challenge is to cause this cycle to end as soon as possible, and then to prevent it from ever starting again.* If we can do that civilization will not only enter the Age of Transition to Sustainability. It will also enter an entirely new mode: a permanent race to the top among politicians, along with all that has to offer, but has never been achieved.

This may seem even more ambitious than the last great political mode change, which was the introduction of democratic forms of government in the 18th century. There is, however, good cause for rational hope, because of:

Chapter 6

The High Leverage Point That Has Never Been Tried

W E HAVE EXTREMELY GOOD NEWS. THERE IS A PROMISING HIGH LEVERAGE POINT in the human system that has not yet been tried. It is <u>general ability to detect political deception,</u> as shown on the revised model on the next page. Pushing there appears to give problem solvers the greatest possible chance of solving the change resistance part of the problem.

Actually the model identifies not one but two high leverage points. Both need their present values raised to solve the problem. But as we will show in another series of simulation runs, it is the key high leverage point of ability to detect deception that makes the biggest difference. (Later chapters expand the model presented in this chapter and identify a third high leverage point. Thus pushing on the high leverage point presented in this chapter is not the complete solution. It is only the beginning of the solution.)

Identifying the correct high leverage points to push on is crucial to problem solving success. But just as important is identifying the low leverage points we should *not* be pushing on. *Environmental activists, academics, politicians, and agencies are failing to solve the global environmental sustainability problem because they are pushing on low leverage instead of high leverage points.* They are doing this because they are using an ad hoc, instinctual problem solving process instead of a formal analytical one, particularly on the sustainability problem as a global whole. If problem solvers would switch to a formal analytical process tailored to the problem, as science did 400 years ago in the 17th century when it adopted the Scientific Method, they would be able to correctly analyze even difficult problems and find the high leverage points necessary to solve them. Only then will the impossible become the possible.

A formal analysis tailored to the problem does not mean find good people, give them the budget they need, apply the Scientific Method, and expect the cows to come home tomorrow. *It means design a custom process that fits the specific problem.*

An example of such a process is the System Improvement Process. Its ten steps are listed on page 169. This process was designed from scratch to solve complex social system problems. It works by breaking the total problem down into ten steps, each of which is a much easier problem to solve. Its main advantages are recognition that change resistance must be overcome before proper coupling can be achieved and the strategy of diagnosis before deciding on treatment. Considering the difficulty of the sustainability problem, these are required advantages.

However, nowhere in environmental activism, academia, political decision making, governmental agencies, or even international bodies have I been able find a group following a process specifically designed to solve the overall

The Two High Leverage Points of the Dueling Loops

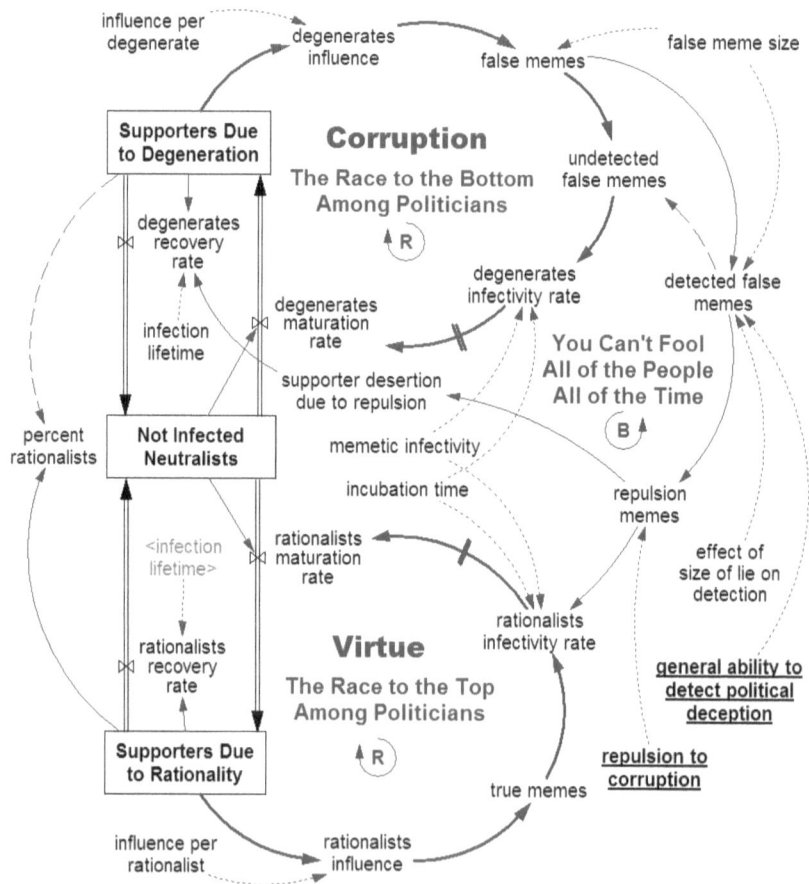

The two high leverage points are underlined. The one making the most difference is general ability to detect political deception. If the model is reasonably correct then pushing there can solve the social side of the sustainability problem. Currently nearly all effort is directed toward the more intuitively attractive low leverage point of "more of the truth," which is the true memes point. Pushing there fails, because environmentalists simply do not have enough force to *directly* overcome the inherent advantage of the race to the bottom. They can only overcome it *indirectly* by pushing elsewhere on high leverage points.

global environmental sustainability problem. This includes the United Nations Environmental Program, the European Union, the Organization for Economic Cooperation and Development, the US EPA, numerous books and papers, and countless NGOs.

What might happen if there was such a group? What if they proved a formal, analytical process tailored to achieving their mission was a better way? Soon there would be a dozen such organizations. What if that in turn caused most environmental organizations to use an appropriate process, either for the complete *problematique* or for the portion of it they were working on?

But we digress. Let's return to the model at hand. On the model a solid arrow indicates a direct relationship. The two dashed arrows show inverse relationships. A dotted arrow is a constant or a lookup table function.

Currently <u>general ability to detect political deception</u> is low. The lower it is the lower <u>detected false memes</u> are. The lower that is, the higher <u>undetected false memes</u> are and the lower <u>repulsion memes</u> are. This causes more degenerates and fewer rationalists, which is bad news.

Currently <u>repulsion to corruption</u> is also low. The lower it is, the lower the <u>rationalists infectivity rate</u> and the lower <u>supporter desertion due to repulsion</u>. This is because <u>repulsion to corruption</u> times <u>detected false memes</u> equals <u>repulsion memes</u>. This makes sense, because detected corruption is a good reason to decide to support virtuous politicians and to desert corrupt ones.

For an actual system reaction to deception detection to occur, two steps must take place. The deception must be detected, which is handled by <u>general ability to detect political deception</u> times <u>false memes</u> equals <u>detected false memes</u>. Then those <u>detected false memes</u> must cause people to be repulsed enough by the corruption to either defect from the degenerates, which is what the <u>supporter desertion due to repulsion</u> variable does, or to become rationalists, which is handled by adding <u>repulsion memes</u> to <u>true memes</u> to calculate the <u>rationalists infectivity rate</u>. In addition to this, <u>false memes</u> minus <u>detected false memes</u> equals <u>undetected false memes</u>, which reduces degenerate infectivity.

Let's summarize how the **You Can't Fool All of the People All of the Time** loop works, focusing on the higher leverage point. Currently the loop is weak, and thus might be more appropriately named **You Can Fool Most of the People Most of the Time**. Low ability to detect deception and the fact that the size of falsehood and corruption can be inflated but the truth cannot combine to cause more supporters to be attracted to the race to the bottom. Thus if ability to detect deception is low, corruption works like a charm, because most <u>false memes</u> flow through the system unimpeded. This causes <u>undetected false memes</u> to be high and <u>detected false memes</u> to be low, which strongly favors the race to the bottom.

But if problem solvers can raise ability to detect deception to a high level, most <u>false memes</u> flow to <u>detected false memes</u>. This greatly decreases <u>undetected false memes</u>, which destroys the power of the race to the bottom. At the same time this increases <u>repulsion memes</u>, which increases the <u>rationalists infectivity rate</u> and increases the <u>degenerates recovery rate</u> due to <u>supporter desertion due to repulsion</u>. The result is corruption doesn't work anymore, which causes the race to the bottom to collapse as most people suddenly see the real truth and flee for their lives to the stock of <u>Supporters Due to Rationality</u>. This is precisely what happens when massive amounts of corruption are suddenly exposed.

It is the effect of influencing so much so strongly that makes <u>general ability to detect political deception</u> such a potent high leverage point.

But even more potent is the fact the dueling loops structure is generic. It applies to any problem, not just environmental sustainability. *The successful exploitation of the race to the bottom by the modern corporation and its allies is the fundamental reason progressive activists are encountering such strong resistance in achieving their objectives.* If progressive philosophy is defined as promotion of the objective truth for the good of all, then progressives (no matter what party they belong to) are rationalists at heart, and thus eschew falsehood and favoritism in its many forms. Progressives may not realize it, but their central strategy is the high road of winning the race to the top.

Next let's familiarize ourselves with how pushing on the two high leverage points affects model behavior. The table below lists the simulation runs needed to do this. In all these runs, the number of initial degenerate and rationalist supporters is 1.

High Leverage Points Model Variables	Simulation Runs							Table 2
	7	8	9	10	11	12	13	14
False meme size	1	1	4.8	4.8	2.4	2	3.8	4.7
Ability to detect deception	0%	20%	20%	20%	20%	20%	60%	80%
Repulsion to corruption	NA	0%	0%	20%	20%	80%	20%	20%
Percent rationalists	50%	100%	0%	41%	20%	57%	69%	100%

Run 7 – This is the same as the reference mode (run 2) presented earlier. The purpose of this run is to test that the revised model has the same foundational behavior. It also serves as a good starting point for further scenarios.

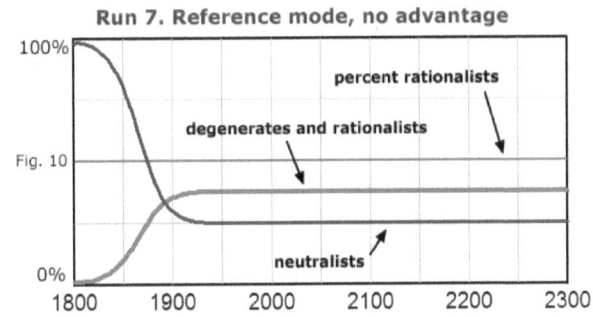

Run 7. Reference mode, no advantage

Fig. 10

Run 8 – In the United States and many other countries, the <u>general ability to detect political deception</u> is low, somewhere around 20% or 30%. This is obvious because of the large amount of political corruption that goes undetected. (A caveat is that recently, in late 2005 in the US, this ability appears to be on the rise due to an excess of corruption that has become intolerable. Sure enough, a corruption cycle event occurred in the US 2006 elections, where many politicians associated with a corrupt party were voted out.) Let's try raising this high leverage point from 0% to 20% and see what happens.

Wow! Great results! Finally it is the degenerates whose graph line is flattened like a pancake. Percent rationalists rises to 75% in 100 years and levels out at 100%. This is a dream scenario. All we've got to do is figure out how to make it happen.

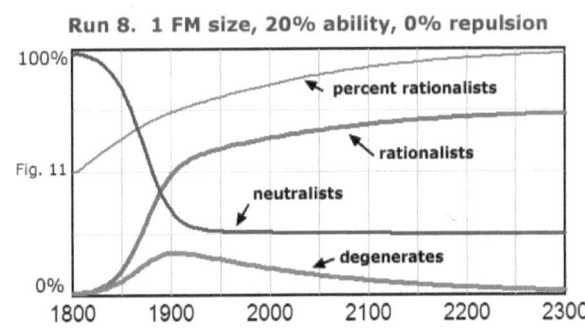

Run 8. 1 FM size, 20% ability, 0% repulsion

Fig. 11

Unfortunately that can't be done, because this scenario is unrealistic. There is no way corrupt politicians are going to sit by and stick to a <u>false meme size</u> of 1, when they know full well, from at least 200,000 years of experience, that corruption works. So let's fix that in the next run.

Run 9 – In this run we change <u>false meme size</u> from 1 to 4.8, which is the optimum size that <u>effect of size of lie on detection</u> and <u>supporter desertion due to repulsion</u> will allow the degenerates to get away with.

Corrupt politicians may be corrupt, but they are not stupid. They are usually expert at adjusting the size of lies and favoritism to be effective without overshoot, which would cause detection. Those unable to do this are quickly selected out by the iron hand of evolution's most merciless law: survival of the fittest.

The graph tells the sad story. Now it is the rationalists who are as flat as a pancake after a *Tyrannosaurus Conservatex* stepped on it. In this scenario they have lost the game so soon and so badly it's as if

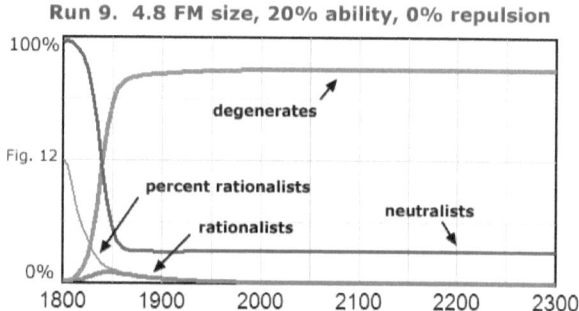

Run 9. 4.8 FM size, 20% ability, 0% repulsion

they had hardly any influence on the political system. But once again, is this a realistic simulation run? Not quite, because repulsion is still 0%, which is unrealistically low. Let's do another run and see what happens when we increase it.

Run 10 – Now we push on the second high leverage point, <u>repulsion to corruption</u>, raising it from 0% to 20%. Because both high leverage points are now being pushed, things should start looking more favorable. If they don't, our understanding of the model is faulty.

The results do look better, but they are still not good enough. <u>Percent rationalists</u> tops out at 41%, which is well below what is needed for a political system to run itself well. We must do better.

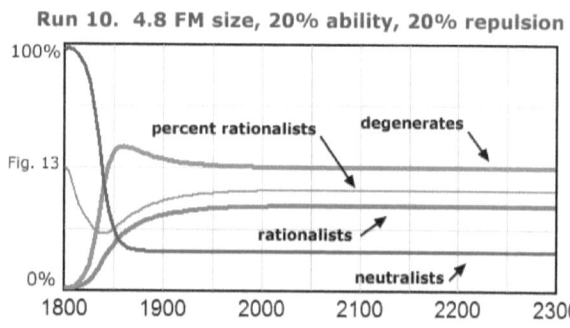

Run 10. 4.8 FM size, 20% ability, 20% repulsion

Run 11 – The smarter the agent, the faster and better it adapts to changing circumstances. We can only assume that corrupt politicians will adapt their strategy to the new circumstances of run 10. Experimentation with the model shows that the optimum <u>false meme size</u> for a 20% ability to detect deception and a 20% repulsion factor is 2.4. So in run 11 let's change <u>false meme size</u> from 4.8 to 2.4.

As the run 11 graph shows, this strategy has a substantially better outcome for the degenerates. <u>Percent rationalists</u> levels off at 20% instead of the 41% of run 10. In other words, the degenerates have increased their percent-

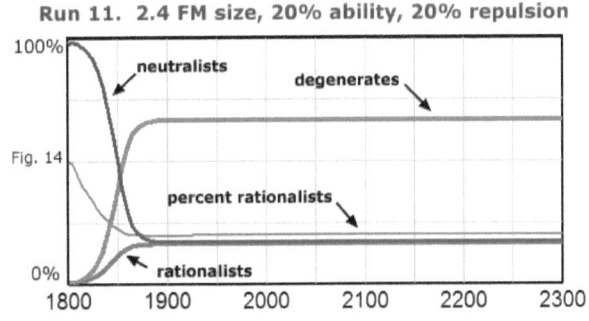

Run 11. 2.4 FM size, 20% ability, 20% repulsion

age from 59% to 80%. Not bad for such a simple change. What's interesting is they did it by *decreasing* the size of lies and favoritism, which means *less* corruption earned them *more* supporters.

The point is that <u>false meme size</u> is not fixed. It is fluid and, like so many agent strategies in complex social systems, changes as the situation demands.

Run 12 – Next let's see which of the two high leverage points gives problem solvers the most leverage. First let's raise <u>repulsion to corruption</u> from low to high, which is from 20% to 80%. Then we experiment with the running model to determine the optimum <u>false meme size</u> for this competitive situation. It turns out to be 2. Will the result be good enough for the rationalists to win or not?

Actually the model is now so complex I found it impossible to reliably predict the outcome of this run. But that's one of the many benefits of simulation modeling: Once you have expressed your analysis as a dynamic structure, the software takes it from there and tells you how that structure will behave in any situation. And unlike my poor overworked cranial lobes, simulation software never makes a mistake.

The results show that even 80% is still not good enough. The forces of truth and corruption are still so evenly matched that they would be totally unable to deal cooperatively and proactively with difficult

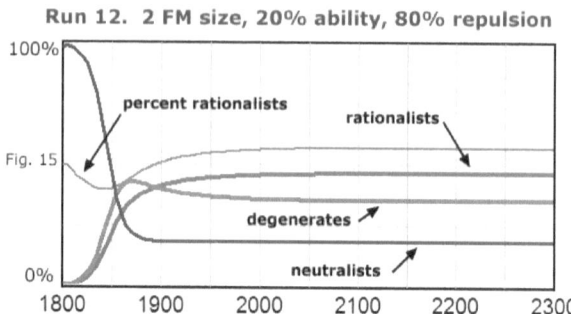

Run 12. 2 FM size, 20% ability, 80% repulsion

Fig. 15

problems, because they would be too busy battling each other. The degenerates would also be engaging in promoting too many wrong priorities to even begin to get behind the right priority of environmental sustainability.

Time for a sanity check. Does this result make sense? Yes, because ability to detect deception is still low, at 20%. So let's roll back repulsion to a more realistic value and then see what would happen if we raised ability to detect deception.

Run 13 – First we must estimate a reasonable value for <u>repulsion to corruption</u>. Later we hope to measure it in the field, but for now we must rely on an estimate.

There are five ballpark values <u>repulsion to corruption</u> could be: zero, low, medium, high, and 100%. Zero and 100% are so extreme as to be unrealistic, so we will rule them out.

I feel that presently <u>repulsion to corruption</u> is low. When the average citizen hears about detected corruption they do very little. They do not take action. Instead, the incident is written off as "politics as usual." Only if corruption is extreme and prolonged do they take effective action. Even when Election Day comes, it is not corruption that voters consider the most. It is numerous other factors, like looks, charisma, sound bytes that stick in the mind, and most importantly, where the candidate stands on issues that are important to each voter. These issues rarely center on corruption, unless corruption has been prolonged and extreme.

Let's not go too low, like 10%. A value of 20% seems reasonable. Much higher would slip into a medium level (40% to 60%), which does not make sense. People do not act on half the corruption they hear about. It is much less.

Also let's start to raise ability to detect deception. In runs 8 to 12 it was 20%. Let's raise it to 60%. Let's continue to assume corrupt politicians will adapt to the new situation and change to the optimum strategy of 3.8 for <u>false meme size</u>. The results are shown below:

This run shows that to adequately counter a false meme size of 3.8, ability to detect deception must be at least 60% and repulsion at least 20%. <u>Percent rationalists</u> is now up to 69%, which is probably about the bare minimum for a

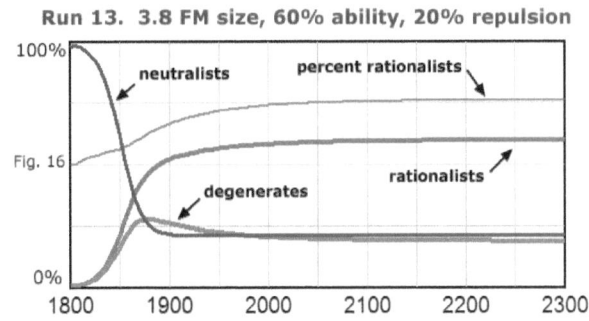

government to begin to put aside political squabbling and begin to work on its backlog of problems. But 69% is still not high enough for nations to focus efficiently on highly demanding problems, because solving these types of problems requires a nation's full attention and its complete cooperation with other nations.

Run 14 – To find out if we can achieve a high enough <u>percent rationalists</u> to solve the problem, let's raise ability to detect deception from 60% to 80%. Again we assume adaptation and change <u>false memes size</u> to 4.7.

The results show that at last we have the behavior in the model we would like to see in the real world, because <u>percent rationalists</u> has risen to a blissful 100%. The opposition is elimi-nated and virtuous poli-ticians can now focus on

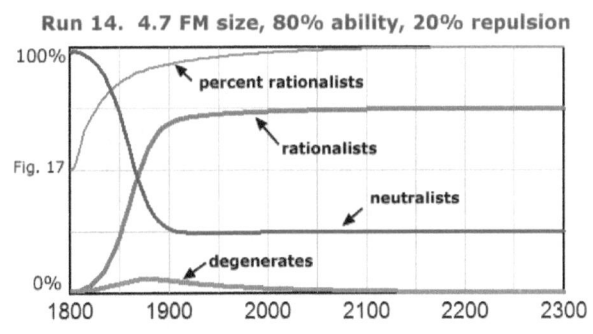

society's proper priorities, at last. *If the model is correct*, then raising the <u>gen-eral ability to detect political deception</u> from low to high is all it will take to make the race to the top go dominant and solve the social side of the problem.

Notice how this run was able to raise <u>percent rationalists</u> from 41% to 100% (a 59% rise) by raising ability to detect deception from 20% to 80%, while run 12 only raised <u>percent rationalists</u> from 41% to 57% (a 16% rise) by raising repulsion from 20% to 80%. Calculating the leverage, 59% / 16% = 3.7. Thus in these fairly realistic scenarios ability to detect deception has 370% more leverage than <u>repulsion to corruption</u>.

* * *

What about leaving ability to detect deception at 60% and raising repulsion to corruption? Would that solve the problem? No. Experimentation with the model shows that increasing repulsion to 80% increases <u>percent rationalists</u> to 94%, and increasing it to 100% only increases percent rationalists to 95%. It seems that increasing repulsion cannot eliminate the last few degenerates. However it does appear that the best overall solution is to raise both high leverage points: repulsion to corruption a little and ability to detect deception a lot.

Now for the important question: *Is the model correct?* No one knows, because it has not been subjected to the rigors of experimental proof and field calibration. However, I do believe that it contains the fundamental brushstrokes explaining why change resistance is so high. At the very least the model should serve as the starting point for a larger project that will go much further.

Next we take up the notion that the dueling loops are cyclic. However, let's first pause for:

A Note of Caution and Hope

At Thwink.org, as well as in this book, we think like scientists. Every assertion we make is a hypothesis that could be overturned tomorrow. The pages you are reading contain many novel hypotheses. While these seem to have withstood the test of logical proof, using a number of analytical tools, few have undergone the acid test of real world experimentation. [38] No one knows how many will survive. But rather than couch every assertion with a "perhaps," a "this suggests," or a "probably," and so on, we have elected to only occasionally stress that all the conclusions in this book are merely examples and pointers to a new way of thwinking. None should be interpreted as *the* analysis or *the* solution.

In particular, the Dueling Loops model itself is only an example. It may or may not be sound. But it should show how, once progressive activists can at last see the social structure of the problems they have been battling for centuries, they will slash right thorough them, with a newfound ease and confidence that will astonish those reading about their adventures many years from now.

Chapter 7

The Cyclic Behavior of the Dueling Loops

U P UNTIL NOW THE MODEL HAS IGNORED CONSIDERATION OF WHAT CAUSES a society to want to raise its <u>general ability to detect political deception</u> and/or <u>repulsion to corruption</u>. To raise the values for these two variables in our simulation runs, all we had to do was reach into the model and change them. That is not how it happens in the real world. How then do societies adjust these values?

My hypothesis is that societies reactively change these values when they see the clear and present need to change them. This need appears when a prolonged excess of corruption occurs. Because there is no formal reliable mechanism to keep the values of these two variables permanently high, they tend to fluctuate as the decades pass. Another way to say this is societies have a short organizational memory on what the values of these two variables should be.

Reactively changing these values causes an endless cycle. This cycle was briefly described earlier: A cycle ends when corruption becomes so extreme and obvious that the people rise up, throw the bums out, and become much harder to deceive for awhile. But as good times return, people become lax, and another cycle begins. These cycles never end, because presently there is no mechanism in the human system to keep ability to detect deception permanently high.

The minimum conditions required for cyclic behavior appear to be:

1. The natural tendency for <u>general ability to detect political deception</u> and <u>repulsion to corruption</u> to be low.

2. The existence of critical points that are automatically activated when corruption gets bad enough. Once a critical point is activated, society invests in raising <u>general ability to detect political deception</u> and/or <u>repulsion to corruption</u>.

3. The critical point is deactivated once corruption falls low enough. This is because there is no permanent mechanism to keep these variables high enough to prevent corruption.

4. The presence of delays in raising and lowering the two variables, and in changing supporters of one type into the other.

The previous model has been revised to incorporate these minimum conditions by renaming the key high leverage point to be <u>Ability to Detect Deception</u> and changing it to a stock instead of a variable. (It is traditional to capitalize the names of stocks, due to their central importance in stock and flow models.) The Critical Point Reaction Subsystem, as shown below, was then built around this stock to give it a realistic critical point and change delay.

A **critical point reaction** occurs when some aspect of a system passes a certain threshold, beyond which the system automatically enters a new behavior mode. For example, water freezes below zero degrees Centigrade. A critical point reaction occurs in political systems when corruption, as measured by <u>percent rationalists</u>, falls below a certain arbitrary cultural <u>corruption critical point</u>.

A corruption cycle works like this: Once the critical point is reached a very common complex social system reaction occurs. The <u>reaction to excessive corruption activated</u> variable goes from false to true, after a <u>reaction delay</u> of 5 years. This causes <u>additional investment</u> to be added to the <u>normal cul-</u>

The Critical Point Reaction Subsystem

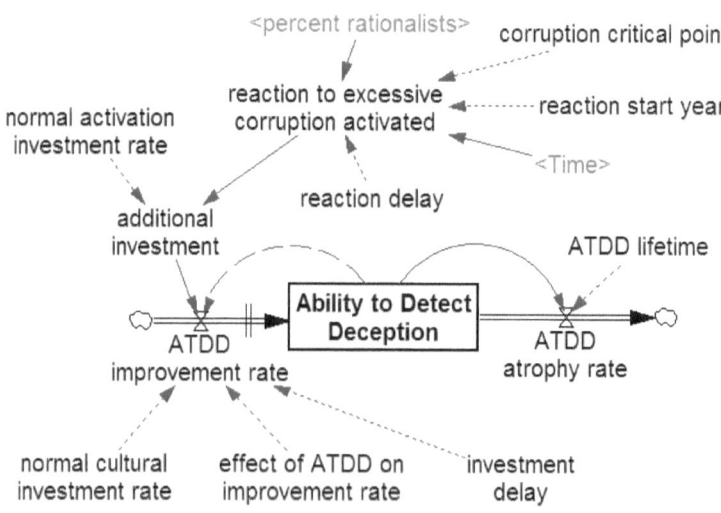

Figure 18. This simple subsystem imitates how society reacts when corruption rises above an unwritten, culturally defined critical point. This reaction is part of a cycle that never ends, because presently there is no formal, enduring mechanism in governments to keep <u>Ability to Detect Deception</u> permanently high.

tural investment rate, which increases a society's investment in raising Ability to Detect Deception, such as by launching investigations, publishing information on who is corrupt, prosecuting corrupt officials, and changing the processes of its governmental institutions to be more corruption proof. This takes time, as represented by the investment delay of 5 years and by the way it takes many years to fill the stock up to the high level needed to detect most corruption.

As the stock of Ability to Detect Deception investments accumulates, more and more false memes are detected. Once the stock rises high enough, so much falsehood and favoritism is detected that percent rationalists rises so high that the corruption critical point is no longer exceeded. This causes reaction to excessive corruption activated to change back to false, which causes additional investment to change back to zero, which causes the stock of Ability to Detect Deception to start falling. It continues to fall until it goes so low that another critical point reaction is triggered, and the cycle starts over again.

Below is the table of simulation runs needed to illustrate the dynamic behavior of the critical point model. In all runs repulsion to corruption is 20%. In a real solution it probably needs to be increased a bit, but here we leave it alone for simplicity.

Critical Point Model Variables	Simulation Runs							Table 3
	15	16	17	18	19	20	21	22
Corruption critical point	0%	35%	35%	50%	50%	70%	95%	100%
False meme size	2.4	2.4	4.7	4.7	5.6	4	4	4.7
Percent rationalists	20%	Very cyclic	40%	Less cyclic	55%	A little cyclic	Barely cyclic	100%

Run 15 – This run has no critical point reaction since the corruption critical point equals 0%. Thus this run's behavior is identical to run 11 because additional investment has not yet been triggered.

The subsystem has a normal cultural investment rate that keeps Ability to Detect Decep-

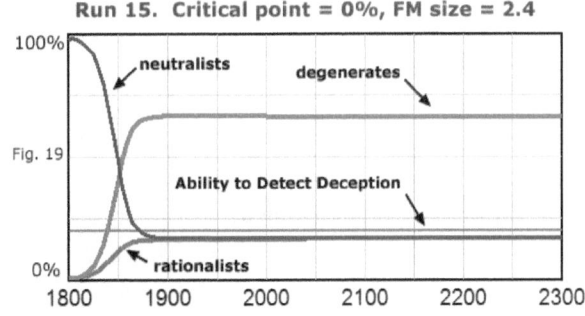

Run 15. Critical point = 0%, FM size = 2.4

Fig. 19

tion at 20% when additional investment is zero. Run 15 is the reference mode for the critical point model. In the graph percent rationalists has been replaced by Ability to Detect Deception, which in this run is a constant 20%.

It takes this run only a hundred years to reach steady state equilibrium. To show the cyclic nature of the dueling loops in later runs, the reaction start year is 1900. Starting the reaction then instead of in 2000 (which would be about now, and make the modeling experience a little more true to life) gives us more cyclic activity to look at, so that we can understand the model and its implications more clearly.

Run 16 – In this run the critical point is changed from 0% to 35%, which means the critical point reaction will take place whenever percent rationalists dips below 35%. Since in the reaction start year of 1900 percent rationalists equals 20%, the critical point reaction starts then. The simulation results show such insightful social system behavior that we have enlarged the graph for this

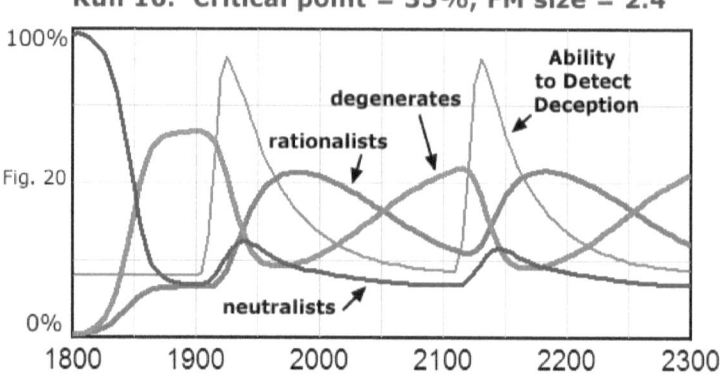

Run 16. Critical point = 35%, FM size = 2.4

run, so the details may be more easily seen.

The graph shows the cycles are about 200 years long. This is much longer than the corruption cycles (really exploitation cycles) we see today. Thus it is more representative of the deeper cycles that occur, such as those due to changes in styles of government, which are a reaction to very deep social system drivers like class oppression by a landed aristocracy or a hereditary line of rulers. If the four delays in the model are reduced to low levels, cycle length falls to about 75 years, which is closer to what we see in cyclic political party dominance or exploitation by life forms or special interest groups like the modern corporation, due to corruption and other related factors that tend to obscure the fact that exploitation of the race to the bottom is the central driver of these cycles. (75 years requires investment delay = 1 year instead of 5, reaction delay = 1 year instead of 5, incubation time = 1 year instead of 10, and infection lifetime = 5 years instead of 20.)

For example, the last time the modern corporation was ruthlessly dominant in the US was in the late 19th century. The cycle was ended with a backlash against the oppressive power of corporations that led to passage of legislation like the Sherman Anti-Trust Act of 1890. But now corporations are overly dominant again, due to successful exploitation of the race to the bottom.

The important thing to realize is that the natural tendency of the dueling loops is to be cyclic. The length of the cycles varies greatly, depending on a host of factors, only a few of which are incorporated in the model. Because there are many corrupt politicians and special interest groups trying to exploit the race to the bottom, there are many cycles underway at the same time. A political system will be most dominated by whichever cycle(s) are currently dominant and by how strong and clever the various exploiters are.

Let's walk through a cycle and explain what's happening, both in the model and the real world it attempts to represent.

A cycle begins when percent rationalists falls below the corruption critical point. Then, after a reaction delay of 5 years we see that Ability to Detect Deception suddenly spikes upward. These spikes are mass panic reactions to flagrant amounts of corruption. When a spike is underway a society will be wildly investing in all sorts of things to increase the public's ability to spot political deception, like editorials and articles explaining how certain politicians are using lies and favoritism to achieve their nefarious goals, investigations to get to the bottom of various scandals and root out corrupt politicians, speeches extolling the importance of virtue and the ravaging effects of corruption, and so forth. Mechanisms to detect falsehood will start spontaneously appearing, such as the way FactCheck.org appeared in the 2004 election in the US, followed by PolitiFact.org in 2007 and TruthFightsBack.com in 2008. (However, efforts like these are not properly focused enough to have more than a modest impact. They are intuitively designed, rather than being analytically designed to push on specific high leverage points. But they are a start, and serve as proof the corruption-to-virtue phase of a cycle is underway and that the US "wants" to raise its Ability to Detect Deception.)

The incubation time of 10 years and other delays causes the percentage of degenerates to not fall as fast or as soon as Ability to Detect Deception spikes upward. Instead, there is a noticeable lag. While it takes only about 25 years for Ability to Detect Deception to reach its peak, it takes about 70 and 80 years for the percentage of degenerates to fall to its lowest level and for the rationalists to reach their peak. These excruciatingly long delays do occur, because it normally takes generations for fundamental cultural norms, like ideology allegiance or addiction to consumptive extravagance, to shift radically.

Once a critical point reaction occurs, eventually the degenerates fall out of power and the rationalists come into power, and a society enters good times. Those times are so good, and what is allowing them is so well hidden, that without realizing it society "forgets" that it should be investing in keeping the Ability to Detect Deception high. The result of this oversight is that very early in the cycle the level of detection ability starts to fall. In this run it starts to fall after only about 25 years, which is 1/8 of the cycle's length. It continues to fall, though the rate of fall slows down as it approaches its normal level of 20%.

In the graph the good times begin when supporter type crossover occurs after about 35 years. After this the rationalists are dominant. This lasts for about half the cycle's length, and then crossover occurs again as the degenerates become dominant. As the percentage of degenerates continues to increase, it eventually triggers another critical point reaction and the cycle starts all over again.

Notice that after 1900 the percentage of neutralists stays within a range of 17% to 29%. This corresponds to the roughly 10% to 30% of the population who are the so called "swing voters." These voters are not strongly committed to either side. If the percentage of rationalists is close to the percentage of degenerates in a political system, as it so often is, then it is the neutralists who determine election outcomes. This fact has not escaped the attention of election strategists.

Run 17 – In the first draft of this model write up I completely missed the fact there is a very successful strategy the degenerates can employ to totally overcome what the rationalists did in run 16. It was only due to correcting a modeling error that I noticed that the wily degenerates have an ace up their sleeve.

Once the cyclic behavior of run 16 begins, the degenerates are dominant a little less than half the time. Thus they are losing. But as the run 17 graph shows, they can win by "losing" even more! This is done by

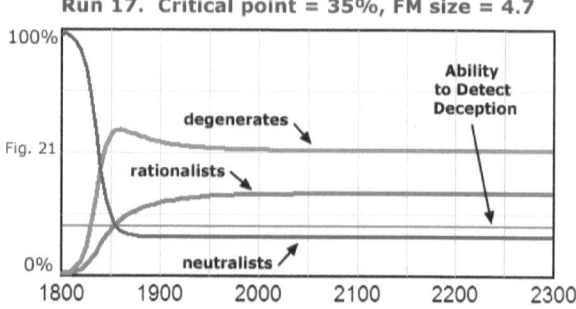

Run 17. Critical point = 35%, FM size = 4.7

Fig. 21

increasing false meme size from 2.4 to 4.7 so as to get caught red handed even more. This causes the pre 1900 portion of the run to level out at 40% instead of the 20% <u>percent rationalists</u> that we saw in run 15. The amazing result is the critical point of 35% <u>percent rationalists</u> is never triggered, the cyclic behavior never happens, and the degenerates, instead of being dominant less than half the time as in run 16, now stay at 60% dominance! How's that for craftiness?

In other words, at a 35% critical point corrupt politicians can win big by telling whoppers they know are going to be detected and cause them to lose more supporters. This corresponds to the flagrant, braggadocio style of lie spinning and cash for favors we sometimes see corrupt politicians or political parties engaging in. There seems to be no logical reason they would try to get caught. But from the viewpoint of the model, there is a perfectly sane reason for such insane behavior: it is the winning strategy. *Figuring out why baffling social behaviors like this occur is impossible without building models like this one.*

Run 18 – It looks like our friends, the virtuous politicians, have no choice but to try a higher critical point. Let's hold <u>false meme size</u> at 4.7 and raise the critical point to 50%.

Once again we have cyclic behavior, though it is a little less so than in run 16. This time the degenerates are dominant only about 10% of the time.

This run begs the intuitive question, if <u>Abil-</u>

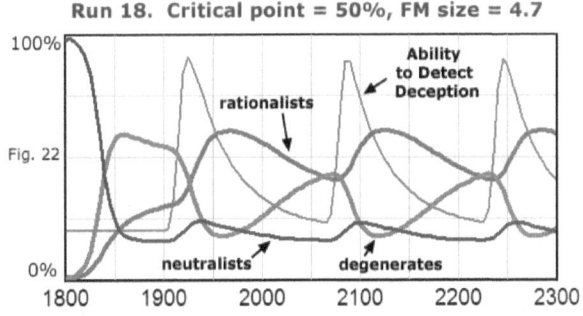

Run 18. Critical point = 50%, FM size = 4.7

Fig. 22

ity to Detect Deception is 50%, then why aren't the rationalists and degenerates each dominant about 50% of the time?

The answer is they would be, if <u>repulsion to corruption</u> was 0% instead of 20%. But 0% is unrealistic, because some people *do* take effective action when they detect corruption, so we have used the value of 20%.

We must not forget for a moment the cleverness of those who believe the end justifies the means. Is there a winning strategy the degenerates can use to counter a critical point of 50%?

Run 19 – Yes there is. Telling even bigger whoppers works like a charm once again. A false meme size of 5.6 allows the degenerates to do much better than being dominant 10% of the time, as in run 18.

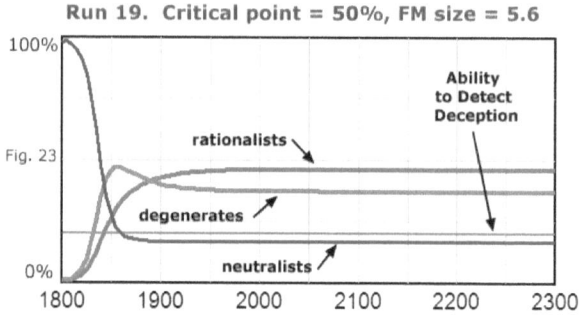

Run 19. Critical point = 50%, FM size = 5.6

Fig. 23

The results show they don't do quite as well as run 18, because now they are in the minority. But they have achieved a dominance of 45%, which is definitely enough to achieve many of their goals, not to mention the sizable impact such a large minority would have on political decision making.

Run 20 –The rationalists need to do much better. Let's get serious and increase the critical point to 70%. Surely this will do the job. At least I hope it does, because raising Ability to Detect Deception that high is not going to be easy.

The results of this experiment are much better, as expected. For the first time the rationalists are safely in control of the political system all the time, by a very comfortable margin. There is still a little

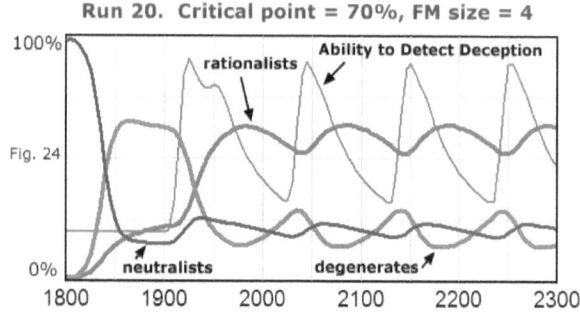

Run 20. Critical point = 70%, FM size = 4

Fig. 24

cyclic behavior, but now the forces of reason are never seriously challenged. The rationalists average about 60% of the population and the degenerates average about 20%.

Once again, is there a strategy the degenerates can use to do better? No. At least not the way this model is constructed. A false meme size of 6.7 does avoid triggering the critical point reaction, but the degenerates average only the same percent dominance. That strategy does not give a better outcome. In this run their best strategy is to maximize their cyclic dominance and use the chaos that causes to try for a lucky victory, which requires adapting to an optimal false meme size of about 4. Thus an important conclusion we can draw from this model is that a high level of Ability to Detect Deception is required to successfully counter the extraordinary power of the race to the bottom.

We are not yet finished. Looking at the graph closely, this run is still not good enough, because even a 20% minority, with occasional swings to over 25%, can still upset the applecart. In modern democracies, every sizable minority still has a voice that must be listened to and frequently accommodated. Thus if a society was trying to deal with a problem so large and difficult that it required all of that society's or a planet's attention to solve it, a 20% minority could prevent that.

So how high does the critical point have to go to solve the problem? That is, how strong does a society's organizational memory have to be for it to always remember how to prevent excess corruption? Let's continue experimenting to find out, by raising the critical point again, this time to 95%. The optimal <u>false meme size</u> of 4 remains the same.

Run 21 – The cyclic behavior is now almost completely gone. But some still exists and there are still a few degenerates to be reckoned with. Is a critical point of 95% good enough to solve problems as intractable as the global environmental sustainability problem?

I think not, for several reasons. One is that as long as some cyclic spikes exist in a social system, it is too easy for those signals to obscure other signals and thus add to the complexity of any problems a society may be trying to solve.

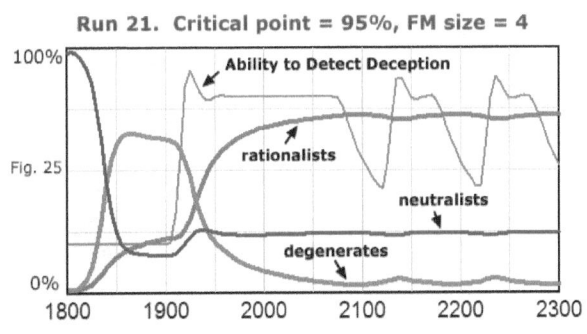

Run 21. Critical point = 95%, FM size = 4

<u>Ability to Detect Deception</u> spikes are not just another signal—they lay at the very heart of human systems, because they are attempts to adjust the perceptual acuity of self-governance. That acuity needs to be at least 20/20 to be able to see the true facts of the many complex, difficult problems governments are responsible for solving. Thus spike signals due to rising degeneration must be responded to in a serious manner, because they may indicate problems of great importance. In addition to the signal confusion problem, spikes in <u>Ability to Detect Deception</u> investment siphon investment away from other endeavors.

There is, however, an even greater reason that a <u>corruption critical point</u> of 95% is not good enough. I believe you can see for yourself what that reason is, from this article that appeared the day after I wrote this. Only the first half of the article is quoted. The rest adds very little to the article's basic argument. (Italics added)

"On Climate Change, a Change of Thinking, by Andrew C. Revkin, The New York Times, December 4, 2005. ~ In December 1997, representatives of most of the world's nations met in Kyoto, Japan, to negotiate a binding agreement to cut emissions of greenhouse gases.

"They succeeded. The Kyoto Protocol was ultimately ratified by 156 countries. It was the first agreement of its kind. But it may also prove to be the last.

"Today, in the middle of new global warming talks in Montreal, there is a sense that the whole idea of global agreements to cut greenhouse gases won't work. A major reason the optimism over Kyoto has eroded so rapidly is that its major requirement - that 38 participating industrialized countries cut their greenhouse emissions below 1990 levels by the year 2012 - was seen as just a first step toward increasingly aggressive cuts.

"But in the years after the protocol was announced, developing countries, including the fast-growing giants China and India, have held firm on their insistence that they would accept no emissions cuts, even though they are likely to be the world's dominant source of greenhouse gases in coming years. Their refusal helped fuel strong opposition to the treaty in the United States Senate and its eventual rejection by President Bush.

"But *the current stalemate* is not just because of the inadequacies of the protocol. It is also a response to the world's ballooning energy appetite, which, largely because of economic growth in China, has exceeded almost everyone's expectations. And there are still no viable alternatives to fossil fuels, the main source of greenhouse gases.

"Then, too, there is a growing recognition of the economic costs incurred by signing on to the Kyoto Protocol. As Prime Minister Tony Blair of Britain, a proponent of emissions targets, said in a statement on Nov. 1: '*The blunt truth about the politics of climate change is that no country will want to sacrifice its economy in order to meet this challenge.*' "

The message I glean from this article is that the solution adoption resistance part of the problem has grown so high that it is no longer just difficult to overcome—it may now be impossible. This is because, as shown in Tony Blair's statement, most of the world is trapped in an *Economic Race to the Bottom among Nations* and doesn't know how to get out. But guess what life form benefits most from that particular downward spiral and therefore has caused it to happen? And guess what high leverage point must be pushed extraordinarily well to stop that downward spiral in its tracks?

The problem is now so close to the threshold of insolvability (or past it, we really don't know) that society no longer has the luxury of tolerating *any* corruption, because any corruption hinders solving the problem and could tip it over the threshold.

One solution alternative is to wait until the first "wake up call" environmental catastrophes start to occur, and then use the belated global realization that humanity *must* solve the problem to move forward on a solution. But if we wait that long, Humpty Dumpty will have already fallen off the wall, and it will not be possible to put all of the pieces back together again.

Why the International Stalemate Exists

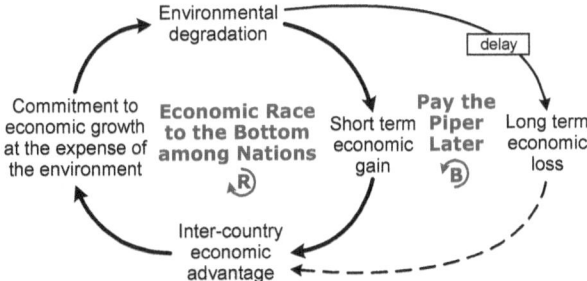

Figure 26. What Tony Blair was really saying is no country can afford to "sacrifice its economy" to get out of the *Economic Race to the Bottom among Nations*. This is because the New Dominant Life Form has structured the international commerce game so that nations see the main loop before the side loop. The way out is to raise ability to detect deception at the level of nations, so they can break free of the illusion they are trapped in the main loop, and can see the truth: that the *Pay the Piper Later* side loop is the more important loop to their citizens.

The main loop starts when a country makes a <u>commitment to economic growth at the expense of the environment</u>. This increases <u>environmental degradation</u>, which in turn raises the <u>short term economic gain</u>, which increases that nation's <u>inter-country economic advantage</u>, and the loop starts all over again, because that is A Good Thing. The side loop shows how, if the *delay* of <u>environmental degradation</u> is considered, then there is a <u>long term economic loss</u> that will eventually decrease the <u>inter-country economic advantage</u>, arguably by much more than the <u>short term economic gain</u>.

The case can even be made that as percent degenerates approaches zero, a multiplier effect is at work. These last few percent are the desperate, hard core degenerates, which includes the smartest of the lot. As percent degenerates goes low, every special interest degenerate ties up two or more for-the-good-of-all rationalists, because (under present conditions) that's how many people it takes to handle damage control and counter the insidious, endlessly disruptive stream of falsehood and favoritism.

Therefore a rule of zero tolerance to political corruption must be adopted, so that *Homo sapiens* is not distracted while it attempts to save itself from ecocide. Anything less is just asking for trouble when it comes to figuring out how to get the US, China, India, and the entire world on board a rapid and radical solution to the climate change problem, as well as to other global environmental problems such as topsoil loss, deforestation, and groundwater depletion.

Let's take a look at what would happen if we tried the rule of zero tolerance in the final simulation run, by using a critical point of 100%.

Run 22 – As expected, zero tolerance to corruption completely ends the cyclic behavior of the dueling loops. Once the rationalists rise to dominance they stay there. Degenerates do not just drop to a low level—they are reduced to 0%. Their best strategy is to hold out as long as possible, by using a <u>false meme size</u> of 4.7. After about 50 years, society's <u>Ability to Detect Deception</u> holds steady at 80%. A successful transition to solving the solution adoption resistance part of the problem has occurred.

But this transition takes a long time. It takes about 25 years for rationalists to begin to outnumber degenerates, and 40 years for <u>percent rationalists</u> to rise to 69% (barely over a 2 to 1 majority), which was mentioned in run 13 as

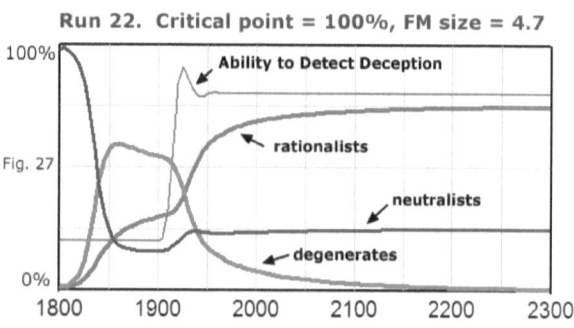

Run 22. Critical point = 100%, FM size = 4.7

Fig. 27

probably the bare minimum it will take to make a serious start on solving the problem, though it is still too low to be enough. As we argued in run 21, it will take somewhere near 100% to be enough.

Because the model is not calibrated (the numbers used in it are estimated, not measured), it cannot make accurate predictions. Nevertheless, it does look as if solving the solution adoption resistance part of the problem will take a long time. Will it take too long? That is one of the great questions facing problem solvers and civilization.

A rare few journalists take the sanctity of truth and zero tolerance to corruption seriously, though because they are unaware of the Dueling Loops, they have only an intuitive sense of why. Well after this book draft was stable for awhile, I ran across this editorial dealing with the deaths of Julia Campbell, budding journalist and Peace Corps volunteer, and David Halberstam, renowned journalist. Here are a few key excerpts: (Italics added)

> "**Working the Truth Beat**, by Bob Herbert, The New York Times, April 20, 2007. ~ I remember once when we were hanging out, shooting the breeze about some horror in the news, Julia said to me, 'Why is the world the way it is?' She added quickly, as though embarrassed: 'I know it's a ridiculous question. But I wonder.'
>
> [David] was among the very best reporters I've ever known. If there was one thing above all else that David taught us, it was to be skeptical of official accounts, *to stay always on guard against the lies, fabrications, half-truths, misrepresentations, exaggerations and all other manifestations of falsehood that are fired at us like machine-gun bullets by government officials and others in high places, often with lethal results.*
>
> " 'You have to keep digging,' he would say, 'keep asking questions, *because otherwise you'll be seduced or brainwashed into the idea that it's somehow a great privilege, an honor, to report the lies they've been feeding you.'*
>
> "*One of the primary tasks of a journalist is to protect the public from such lies by exposing them, and by reporting the truth.* David Halberstam was a master at that.
>
> "In a larger sense, our job has to do with the question Julia Campbell asked in those days when her heart was set on a career in journalism. *We don't know why the world is the way it is,* but the job of the journalist is always in some sense to chase after the answer to that question."

If Bob Herbert, David Halberstam, and Julia Campbell had known about the Dueling Loops, they would know why the world is the way it is, when it comes to the lies, fabrications, half-truths, misrepresentations, exaggerations, and other manifestations of falsehood that are fired at them like machine gun bullets by politicians who are trapped in a race to the bottom.

With that knowledge, journalism might come to see why "reporting the truth" is so crucial. It might also see that due to the inherent advantage of the race to the bottom, it is not enough for journalists to "protect the public from lies by exposing them and by reporting the truth." That is a myth. It will not work, because journalists "working the truth beat" control only a small fraction of the political meme stream, one growing smaller each year. "Reporting the truth," unless approached in a comprehensive manner as the next chapter attempts to do, is also the same as pushing on the low leverage point of "more of the truth." It simply will not work. And it *has* not worked.

This completes the presentation of the basic dueling loops simulation model. A later chapter in this book, The Battle for Niche Succession, extends the model into its full form.

Does the Dueling Loops Really Exist?

We don't know yet. The model is so new it has not yet been rigorously tested. But there is a way for you to perform some of that testing yourself. If you like what you see you can become an early adopter, long before the herd.

James Trefil's *The Nature of Science*, 2003, is a 400 page "Guide to the Laws and Principles Governing Our Universe." The book devotes several pages to each of the two hundred key "laws of nature." Taken together:

> "The laws of nature are the skeleton of the universe. In an age that seems to be losing confidence of its ability to manage things, [the laws of nature] remind us that even the most complex systems around us operate according to simple laws, laws easily accessible to the average person." (page vii)

Laws explain regularities. Trefil cautions that "a law is just as likely to be known as a theory, a rule, a model, or a principle, [or even] a relation or an equation," such as the *theory* of evolution, Newton's three *laws* of motion, Hamilton's *rule*, or the Copernican *model* of planetary motion. How are such laws discovered? According to Trefil: (Italics added)

> "One of the great truths that we have discovered is that *we live in an ordered universe*, a universe whose workings are accessible to the human mind. The enterprise we call science differs from other attempts to interpret the universe in that it does not seek absolute truth, but instead uses a method that produces *successively better representations of physical reality*. The Scientific Method begins with a question: Why do things happen this way and not some other way? The scientist explores, *systematically observing and measuring*, looking

for correlations and anomalies. *Once a pattern emerges, an explanation is framed.* The more general the explanation, the more *predictions* it will make about how other things should happen. The scientist continues to observe and measure, to test those predictions. *If the explanation survives those tests, the result is a law of nature."* (page xxix)

Construction of the Dueling Loops model began with the question: Why does the human system resist change that is "better" for the system as a whole? Why does the system behave that way and not some other way?

Several years of systematic observation of the history of the sustainability problem led to discovery of the pattern that the more corrupt a political system is, the less it addresses important problems like sustainability. Once that pattern emerged, an explanation was needed. First the race to the bottom loop appeared. This explained why common good problems were deemphasized, so that special interest problems could be favored instead. Later came the race to the top loop, to explain why the system sometimes tilted toward solving common good problems. Further observations and model testing led to refining the model so that it better explained why the system behaves the way it does. At this point the model stabilized.

Study of model behavior has led to eight key predictions:

1. As falsehood is increased the race to the bottom becomes more dominant, as indicated by more corruption and a preference for solving problems important to special interests.

2. There is a point of diminishing returns in the size of falsehoods and favoritism.

3. Those citizens who support virtuous politicians will tend to push on the leverage point of more of the truth, since it's the obvious way to counter falsehoods and shift to solving common good problems.

4. This fails to work because it's a low leverage point.

5. If virtuous supporters would push on the high leverage point of raising general ability to detect political deception instead, the system would shift modes to a dominant race to the top. This mode would exhibit low corruption and a preference for solving problems of a common good nature.

6. General ability to detect political deception is normally low.

7. Since there is nothing in the system to keep it high, the system will exhibit cyclic behavior.

8. Permanent dominance of the race to the top can be achieved by pushing on the high leverage point of quality of political decision making. (This is presented in a later chapter.)

This is a rich collection of useful predictions. They can be tested two main ways: First, do they agree with past behavior of the system? Second, do they accurately predict future behavior, either when experiments are run or when the actual system runs?

Over the past four years I have informally tested these predictions, mostly by observation of the past. All hold up very well except number 5 and 8, which have never been tried. I've sprinkled some of these observations throughout this book, in an effort to build a solid case for the validity of the model as well as to explain how it works.

But don't take my word for it. *You can test all the predictions but number 5 and 8 yourself in a few minutes.* Read the paragraph containing them again. As you come to each prediction, ask yourself: Has this been happening? Is the historic pattern uniform enough to say this is true?

If most thoughtful readers answer yes, then the Dueling Loops probably exist. If they continue to say yes for the next few decades and a better model does not appear, then we have new natural law.

We can also test the model by making forward looking predictions. *The model hypothesizes that corruption is cyclic.* As a prime example, examine the cycle underway in the United States. Corruption was low in the Bill Clinton years. It grew high during the George W. Bush administration of 2001 to 2008. Towards the end of that period corruption and its consequences grew so bad that the press and the people became alarmed. (At this point I was able to start making rough predictions that came to pass, since the model stabilized in late 2005.) In the election of 2006 Americans began to take action and voted in many virtuous politicians, despite heavy use of political deception by those in office. This trend continued in the election of 2008, where voters have become so irate at the Bush administration that an anti-Republican landslide was bound to happen. It has, though it has been helped by the financial meltdown. But this too was caused by corruption, which led to lax regulation of financial institutions and unsustainable economic management, which led to the subprime mortgage crisis, which triggered the crisis.

The model predicts that once virtuous politicians are in office the system will lean toward solving common good problems for awhile. But over time this mode will degrade, first to a state of mixing common good and special interest goals, and then to a corrupt state of mostly special interest goals. Once that occurs, another cycle will be complete, and we will have further proof that the Dueling Loops are driving the behavior of the system.

There is a window of opportunity here. If we can take advantage of lower change resistance while the system is in the virtuous state, then we can try to accomplish two things: We can properly couple the human system to the environment to solve the sustainability problem. But that solution (and many others) will not hold unless we accomplish a second task: changing the system so that it naturally wants to stay in this state. *The model predicts this can be done by pushing on the high leverage point of quality of political decision making.*

Then we will have broken the cycle. We will have achieved what the inventors of democracy envisioned long ago: a system that works all the time for the common good of all.

We will have also proven the model exists, and science will have advanced one more step. By then the model will have evolved considerably.

Summary of the Diagnosis of the Root Cause

At the beginning of this book we promised to diagnose why progressives are stymied. The reasons are subtle. Finding them requires the structural thinking tool of modeling and the use of a process tailored to the problem type, such as the System Improvement Process.

The top long term problem facing progressives is the global environmental sustainability problem. This has been used as a running example of how the paradox can be solved. If we can solve this problem then we can probably solve them all, because they are all complex social system problems, and they all appear to be the result of exploitation of the race to the bottom.

By going beyond the *technical side* of the sustainability problem to the *social side*, which is the crux of the problem, we arrived at the dueling loops model. This consists of **The Race to the Bottom among Politicians** battling against the race to the top for the same supporters. Whichever loop can offer uncommitted supporters the most perceived benefits wins.

The race to the bottom has an inherent structural advantage over the race to the top. This causes the race to the bottom to be dominant most of the time. Because the race to the bottom requires generous amounts of falsehood and favoritism to work, that is what characterizes politics today.

The modern corporation and its allies is the New Dominant Life Form. Because it is the dominant special interest, it controls the race to the bottom, and thus the political systems in industrialized countries. It doesn't control all of each system, but it controls enough to cause the rules of the game to be defined in its favor. It also controls enough to acquire the favoritism needed to remain dominant. In this manner the modern corporation has become human-ity's master, and most of us it's compliant, uncomplaining ideoserfs. *An* **ide-oserf** *is someone who is bound to an ideology, as serfs were bound to the land.* An ideoserf is also called an *incognizant proxy.*

Corporations are each in their own life or death struggle, based on who does the best at maximizing the net present value of profits. This causes the life form as a whole to be locked into a preference for unsustainable behavior. Because corporations are the dominant life form, this in turn causes the entire system to be locked into the mode of unsustainability and unprogressiveness.

Thus the root cause of the Progressive Paradox is a dominant race to the bottom, due to exploitation of that loop by the New Dominant Life Form.

Hostile and successful opposition to progressive ideals is an emergent property of the structure of the system. Because the structure of the human system is largely invisible, most problem solvers have responded by pushing on an inviting but low leverage point. This is to spread as much truth as possi-ble about progressive problems, and hope that people will see why solving them proactively is in their own best interests.

This solution, known as "more of the truth," has become the *modus oper-andi* of the progressive movement, and is thus the only solution the movement has. It works on easy problems but fails on the difficult ones, which includes the most urgent problem of them all: climate change. Despite repeated failure, different versions of this solution keep reappearing ad infinitum, *because progressives have no other solutions.*

A key finding of the analysis is that "more of the truth" is a low leverage point. *Pushing on this point fails because it is no more than a heavy handed, naive attempt to make the race to the top dominant through the application of brute force.* It does not consider that the race to the bottom is inherently stronger and has a more powerful special interest group behind it. Thus con-ventional solutions have no hope of succeeding, unless the laws of physics change or a "wakeup call catastrophe" occurs in time. Neither appears likely.

Fortunately there is at least one way out: the high leverage point of gen-eral ability to detect political deception. Currently this is low. If problem solv-ers can raise it to a high level the race to the bottom will collapse, causing the race to the top to go dominant. Politicians will then respond to the truth about the global environmental sustainability problem because it will now be in their best interests. If they come to the same conclusion that environmentalists have, that sustainability is civilization's top priority and nothing else comes close, then civilization will at long last enter the Age of Transition to Sustain-ability.

Chapter 8

How to Raise the Ability to Detect Political Deception

WHAT WE ARE ABOUT TO PRESENT MAY SOUND HOPE-
LESSLY NAÏVE. At first glance it may appear there is no earthly way
it could work. Indeed, this is the way people reacted at first to Jay Forrester's
analysis of the urban decay problem:

> "The conclusions of our work were not easily accepted. I recall one
> full professor of social science in our fine institution at MIT coming
> to me and saying, *'I don't care whether you're right or wrong, the re-*
> *sults are unacceptable.'* So much for academic objectivity! Others,
> probably believing the same thing, put it more cautiously as, *'It does-*
> *n't make any difference whether you're right or wrong, urban officials*
> *and the residents of the inner city will never accept those ideas.'* " [39]

What is really happening here, at the deepest appropriate level of abstraction? The 19th century German philosopher Arthur Schopenhauer knew exactly. He put it this way, in what has become one of the best known quotes in
the advancement of science:

> *"All truth passes through three stages. First, it is ridiculed. Second, it*
> *is violently opposed. Third, it is accepted as self-evident."*

For example, an "intelligent, articulate" man from Harlem in one of Forrester's educational seminars passed through these three stages in a matter of
days. On Monday he ridiculed what Jay Forrester was presenting when he
said, "I come from Harlem and there's certainly not too much housing in Harlem." Next, one of Forrester's students reported on Tuesday evening that, "the
group was very hostile." At that point the man was in the second stage. Four
days later when he said to Forrester, "You know, it's not a race problem in
New York at all, it's an economic problem," he had reached the third stage. He
had accepted the full truth of the model of urban dynamics, along with its
counterintuitive but undeniable conclusions. [40]

I now ask you to put yourself in that man's shoes, because the truth that is
about to be presented may be just as unacceptable—at first.

The truth is that if experimental confirmation shows the Dueling Loops
model to be sound, then the solution elements presented later in this chapter

have a high probability of solving the problem, however unconventional and counterintuitive they may appear to be.

However, this is part of an even greater truth, a greater conceptual whole. This is the causal chain that leads from problem discovery to successful solution when the proper problem solving process is applied. How this looks is shown below:

The Complete Problematique Chain

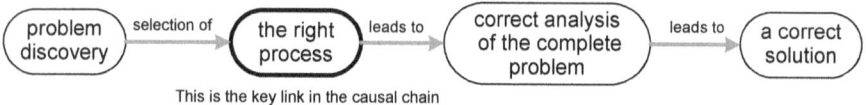

This is the key link in the causal chain

Emphasis on **the complete problematique** is a systems thinking concept promoted by Aurelio Peccei, an Italian industrialist who founded the Club of Rome in 1968. His point was that for the incredibly complex and interlocking problems global society now faces, only a sufficiently *complete* analysis of the meta-problem can realistically expect to solve any of the subproblems. [41]

The chain holds only if each link in it is strong. As Aurelio Peccei so presciently observed, a correct solution can only follow a correct analysis of the *complete* problem.

Let's briefly review the process presented and used in our approach to solving the sustainability problem. The System Improvement Process has these four main steps:

1. Problem definition

2. System understanding

3. Solution convergence

4. Implementation

The chain starts with problem discovery. Unless it is a simple problem, the next step *must* be selection of the right process. Applying the right process leads to correct analysis of the complete problem, which is steps one and two of the System Improvement Process. If this is done well, then the analysis leads to a correct solution, which is steps three and four of the process.

If this chain is conceptually sound, then failure to solve the problem can only be due to one or more weak links in the chain. Failure to solve the problem has clearly occurred. This forces us to ask: Which link or links in the chain are weak?

My conclusion is that the second link in the chain, the right process, is the culprit. My reasoning on why this is so runs like this:

In 1972 the international bestseller *Limits to Growth* brought the global environmental sustainability problem to the world's attention. This and other events spawned the modern environmental movement. Since then millions of environmentalists have relentlessly attempted to solve the problem. Some success has occurred. But this has been only on the easy problems, the low hanging fruit. The more difficult problems, which are the ones where solution adoption resistance is strong, remain unsolved.

THE LIMITS TO growth

Donella H. Meadows
Dennis L. Meadows
Jørgen Randers
William W. Behrens III

A Report for THE CLUB OF ROME'S Project on the Predicament of Mankind

A POTOMAC ASSOCIATES BOOK $ 2.75

Here is a hypothesis for why this happened: *Limits to Growth* (as well as most other efforts) analyzed only the *technical side* of the sustainability problem. By modeling only the environmental, economic, demographic, and technology aspects, it left out the *social side* of the problem. *This is the crux of the problem.* In general, society knows what it must do: live sustainably, which is the technical side. But for rational reasons many powerful agents refuse to do so. This causes *change resistance,* which is the social side of the problem.

By omitting consideration of the change resistance part of the problem, *Limits to Growth* implied this was not necessary. Due to the influence of the book in framing the debate over the next several decades, this fateful omission steered problem solvers away from what has turned out to be the crux of the problem. But this should not detract from the vital contribution the book made, which was to correctly identify the sustainability problem for the first time.

The result is that now, 35 years later, it appears no one has addressed the social side of the problem successfully, because the processes used (particularly Classic Activism, have not gone far enough beyond the analysis introduced in *Limits to Growth* in 1972. *Thus the next step is to use a process that includes the social side,* such as the System Improvement Process. This will allow us to tackle the complete *problematique,* in a manner comprehensive and mature enough to solve it.

This leads to the most fundamental truth of them all. It is the one that environmentalists *must* accept fully, if they are to improve their operative model and have any rational hope of solving the problem in time. *This is the critical importance of using the right problem solving process.*

Therefore if the modern environmental movement wants to succeed, it must acknowledge this new truth, and build the proper second and third links in the chain. This will lead to a strong fourth link, which is the real goal of the chain. Doing a good job of this will probably require a collective effort, such as a coalition of leading environmental organizations, because of the large amount of investment, experimentation, coordination of effort, and expertise required. Or perhaps one bold organization will lead the way.

The right process link is the key link, because if it is strong, then the chain will hold. But if it is weak the chain will usually not hold, because the next link will usually not be a correct analysis. That is exactly what has happened here, and therefore the chain is broken. The result is the complete *problematique* has never been fully and correctly addressed.

Applying the Right Process

The right process, as Jay Forrester and so many others have shown, is a process with the right steps and the right tools for the problem at hand. If problem solvers take the wrong steps and use the wrong tools, then no matter how hard and long they try, a truly difficult problem will not yield to even heroic efforts except by luck. That occurs so seldom that it would be more than a little irresponsible to bet the future of *Homo sapiens* on the wrong process.

This book is a modest demonstration of what happens when the right process is applied to the global environmental sustainability problem. To maximize the chance of solving this problem, as well as the other complex social system problems entangled with it, in 2001 when I began work on this problem I paused and took the time to design an appropriate process from scratch. This is the System Improvement Process. It has four simple steps.

The first step is Problem Definition. The second step is System Understanding. This is where problem solvers should spend about 80% of their time. *If the all important second step is done well, problem solvers (and anyone else, including decision makers) will understand the system with the problem so deeply and correctly that the third step, Solution Convergence, is almost trivial.* Problem solvers will understand the dynamic structure of the system so completely that they can predict, within a broad range, how it will respond when low, medium, and *high leverage points* are pushed on. Solution Convergence then becomes a simple matter of selecting a reasonably straightforward way to push on the high leverage points. Because the correct points will be used, almost any form of pushing on them will do. *A seemingly trivial solution is the payoff for using the right problem solving process.*

We have concluded that the general ability to detect political deception was the key high leverage point. If problem solvers can raise it to a high level, then the race to the bottom among politicians will collapse, leaving the race to the top dominant. Politicians will now be competing on the basis of who can provide the most benefits to society as a whole, based

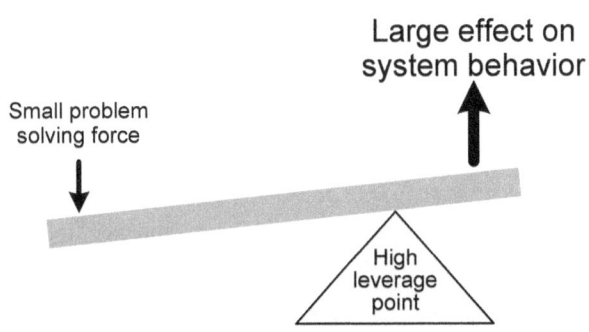

Large effect on system behavior

Small problem solving force

High leverage point

The choice of the right **high leverage point** (HLP) allows a small problem solving force (the total effort required to prepare and make a change) to have a large effect on system behavior. This requires choosing the right change force and the right application point. In a complex social system, leverage is the use of *indirect* force rather than *direct* force. The highest leverage is achieved by pushing on HLPs such that feedback loop dominance changes radically. This requires seeing the social structure involved, so that the right HLPs are used and are pushed on correctly.

on the objective truth. It will not take them long to realize that their top priority needs to be global environmental sustainability, causing that problem to finally receive the full attention and commitment it deserves.

But that will never happen unless the general ability to detect political deception can be raised from low to high.

The Solution Convergence step of the System Improvement Process has discovered that it takes six solution elements to do this. The first is the foundation for all the rest. It is:

The Freedom from Falsehood Solution Element

Hindsight sharpens the vision. Most difficult social problems have, in retrospect, what appears to be a surprisingly simple solution. Looking back at history, it almost seems the bigger the problem, the simpler the solution. For example, the Magna Carta of 1215 introduced the idea that a ruler's subjects have rights that must be respected by law. The invention of democracy gave a population the right to choose its own leaders, who must respect the population's lawful rights. The ending of serfdom and slavery gave serfs and slaves the right to freedom from control by their former masters. Each of these solutions solved an age old, seemingly intractable problem with a solution so simple that we can now describe it in a single sentence.[42]

Civilization remains saddled with a problem that is every bit as debilitating and exploitive as any problem the solutions above solved. Ever since politics began, corruption has been the norm. Corruption is so rampant that a "good" politician is not the one Diogenes could hold a lamp up to and say, "This is an honest man." Instead, a good politician is one who is the least corrupt. That we are forced to choose from the lesser of the evils is pathetic and perverse. [43]

But this need not be so. Diogenes would find an honest politician every time he held up his lamp if people had the right to freedom from falsehood.

Freedom from falsehood gives people the right to freedom from falsehood from sources they must be able to trust. This includes all "servants" of the people, such as politicians, public employees, and corporations. A **servant** is an agent created or employed by *Homo sapiens* to do something useful. All servants must remain subservient to *Homo sapiens* and keep the interests of humans above their own.

What is not prohibited by law is permitted by implication. Therefore if people do not have the legal right to freedom from falsehood, then by implication it is okay for those in positions of power to manipulate citizens by the use of lies, fallacies, the sin of omission, and all the forms of deception, propaganda, and thought control available.

Corruption relies on the use of falsehood to hide or rationalize favoritism. Eliminate falsehood, and you have eliminated favoritism. This is because once falsehood is banished, politicians will be forced to compete for supporters on the basis of the objective truth. The truth includes the long term optimization of the general welfare of all members of *Homo sapiens*. Favoritism conflicts with this goal because it gives someone more than his or her fair share, and hence someone else less. This promotes the welfare of an elite few, rather than that of the many, so it is not the optimal allocation of a society's resources.

If "we the people" do not have freedom from falsehood, then falsehood in all its Machiavellian and Orwellian forms will continue to appear again and again, because it is the surest way to *rise to* power, *increase* power, and *stay* in power.

Activists are intuitively coming to the conclusion that freedom from falsehood is essential. As one example, in an article on May 15, 2007 Julian

Burnside, a prominent Australian barrister, advocated almost exactly that. Here's the beginning of the article: (Bolding added)

> "The Future Summit, being held in Melbourne this week, is a hotbed of ideas, solutions and attempts to imagine a better world.
>
> "Global warming, reliance on fossil fuels, the growing gap between rich and poor, all have been debated by academics, captains of industry, religious, community and political leaders.
>
> "But one solution — put forward yesterday by the top silk Julian Burnside, QC — met with more acclaim than any other, and received rapturous applause.
>
> " **'If we really want to make things better, I suggest we introduce a law that makes it an offence for politicians to lie,'** he told the conference." [44]

Julian Burnside has intuitively sensed what the Dueling Loops model analytically shows: that political deception is so damaging to democracy it should be illegal. The way to make that happen is to recognize that as long as the democratic model lacks the fundamental right to Freedom from Falsehood, it is an incomplete and too easily compromised model.

However this new right alone will do little good unless falsehood can be detected. This is why we need:

The Truth Test Solution Element

The Truth Test is a personal skill, much like other skills such as frugality, language, and mathematics. It is designed to handle nearly all arguments the average person receives in seconds or minutes. The rest take longer or an expert.

The objective of the Truth Test is to reduce deception success at the individual level to a very low, acceptable amount. It consists of four simple questions:

1. What is the argument?
2. Are any common fallacies present?
3. Are the premises true, complete, and relevant?
4. Does each conclusion follow from its premises?

The Truth Test allows people to see the widespread fallaciousness of the arguments they receive from corporate proxies, such as corrupt politicians, many news sources, and articles. Once citizens can no longer be fooled by unsound arguments, they will elect better leaders and support better positions.

We certainly don't expect the general population to master the Truth Test very soon. But we do expect those performing Truth Ratings (described below) to do so, as well as those who are trying for high Truth Ratings.

As the general population sees the published Truth Ratings and occasionally reads the details behind a rating they are particularly interested in, they will get a long, gradual exposure to how the Truth Test works. This and more direct educational efforts will gradually lead to **truth literacy**, which is the ability to tell truth from falsehood.

Universal truth literacy is just as important to society as reading literacy, because if people cannot "read" the truth, then they are blind to what the truth really is.

The average person is never taught anything like the Truth Test

> For a complete introduction to the Truth Test, see the *Truth or Deception* pamphlet and video at Thwink.org. The pamphlet runs 48 pages, the same length as Thomas Paine's *Common Sense*. Both were designed for the same purpose: to wake up a nation to a new fundamental truth.
>
> As this manuscript was entering its final edit in October 2008, Robert Gowans of Sweden was just starting to setup TruthTest.org, a site dedicated to implementing the Truth Test solution element.

in school or the workplace. *Thus their immunity to deception is largely a matter of cultural chance.* For truth literacy to become a cultural norm and achieve its full success, it must become as essential to a person's education as reading and writing.

History has shown again and again that those who are not truth literate become the unknowing slaves (really ideoserfs) of the masters of falsehood, as the cyclic nature of the race to the bottom versus the race to the top plays itself out over and over. A cycle ends when corruption becomes so extreme and obvious that the people rise up, throw the bums out, and become much harder to deceive for awhile. But as good times return, people become lax, and another cycle begins. These cycles never end, because presently there is no mechanism in the human system to keep ability to detect deception permanently high.

The appalling effects of this cycle, during which corrupt politicians and special interests are dominant most of the time, is historic evidence that truth literacy is more important to society than reading literacy. This applies even more so today as we enter the 21st century, because if the truth is not seen in time, *Homo sapiens* will surely perish by his own hand.

How the Truth Test Works Dynamically

Implemented properly, the Truth Test is true structural change. It works by introducing the reinforcing feedback loop shown below:

Once a person completes initial study of the Truth Test the cycle of *Lifting the Blanket of Deception* can begin. Use of the Truth Test increases the amount of falsehood spotted on everyday arguments. This increases quality of decisions. Once a person perceives this has happened, an increase in knowing you benefited from better decisions occurs. This causes that person to use the Truth Test even more, and the main loop starts over again.

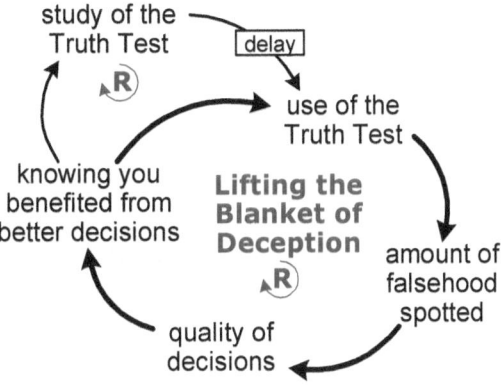

The Dynamic Structure of the Truth Test

The Truth Test lifts the blanket of deception higher and higher by the more you use the Truth Test, the more you benefit, and so the more you want to use it.

Let's examine the side loop. Knowing you benefited from better decisions will increase study of the Truth Test. This occurs when people realize that if they study the test more, they can handle a broader range of arguments and make better analyses. Or there may be a particular type of argument they would like to handle better. After the *delay* of learning, there will be a tendency to use the test more, because now it can offer them even greater benefits.

Nothing can grow forever, so these reinforcing loops have balancing loops associated with them. Examples are the increased time and cost of using the test, and the increased complexity or cleverness of arguments. Each of these causes diminishing returns, which keeps the *Lifting the Blanket of Deception* loop from growing forever. For simplicity these additional loops are not shown.

As just one example of how the Truth Test might affect society, imagine what a talk show might be like if the host was trained in the Truth Test and was familiar with Truth Ratings. After a particularly fallacious string of comments from a guest, such as one from a biased think tank, the host might reply with "By the way, while you and I have been talking, my assistant was jotting down how many fallacies and truths you uttered, and what kind. Did you realize that since you began ten minutes ago, out of a total of 24 propositions, 6 were *ad hominem* attacks, 4 were based on biased samples, and 8 were false enemies or pushing the fear hot button without any justification? This leaves only 6 reasonably true propositions. In other words, in my opinion your sequacious punditry is false 75% of the time. THAT is the real news here. And…, let me see, my assistant reminds me that it was about the same last time you were on. What do you say to that?"

The silence that followed might be the sound of the beginning of the race to the top.

The Truth Test provides a way for citizens of all kinds, including talk show hosts, to spot the truth. But it is a bit of a stretch to expect that truth literacy will sweep the world soon. The Truth Test also provides no irresistible incentive for corrupt politicians to start telling the truth. For that we need:

The Truth Ratings Solution Element

Truth ratings would provide an accurate measure of the truth of what key politicians are saying and writing. If this objective can be achieved, then construction of a new reinforcing loop causing virtue to triumph over corruption in the political arena becomes possible. *Once this new loop is established, it become increasingly difficult for political deception to succeed.*

Truth ratings work by rating the truth of important statements made by important politicians. They are similar to other types of ratings that have been around for a long time.

Credit ratings quantify the creditworthiness of a person, organization, or government. Product ratings, such as those in Consumer Reports magazine, quantify the worthiness of products. Both are widely used. Truth ratings would quantify the truthfulness of important arguments, such as those in political statements, articles, and so on.

A **truth rating** is the probability an argument is true. For example a few days after a presidential debate, its truth ratings would come out. They might say that candidate A averaged 45% true, while candidate B averaged 70%. Guess which candidate would probably win the debate in the public's mind?

If the organization doing the rating was credible and the public trusted the truth ratings, a race to the top would begin. Politicians would compete to see who could be the most truthful in the fullest sense of the word, and therefore the most helpful. Campaigns would become based on reason and truth rather than rhetoric. Due to a trickle down effect from the successful use of Truth Ratings, a race to the top would also begin in many other areas of society where less than the truth has long prevailed, such as advertising, the appeals of special interest groups, editorials, and to a growing degree, the news.

No one person can become an expert on the many critical issues of our day and spend hundreds and sometimes thousands of hours analyzing each important political argument they encounter. Therefore the public has no choice but something like Truth Ratings.

Instead of individuals continuing the impossible task of deciding the truth of each important argument, rating organizations would do that. Certified rating organizations would *quantify* the truthfulness of important arguments by applying the Truth Test and providing a written rationale for each rating, so that the public could make its own final judgment. As they read more about the logic behind ratings of interest, the public would gradually become educated in how to apply the Truth Test.

Efforts to provide the beginnings of truth ratings are springing up spontaneously. For example, in October of 2006 Eric Schmidt, chairman and CEO of Google predicted:

> "...that, within five years, 'truth predictor' software would 'hold politicians to account.' Voters would be able to check the probability that apparently factual statements by politicians were actually correct, using programs that automatically compared claims with historic data." [45]

Politicians are not the only social agent needing truth ratings. Another is the news media, where fiction is too often presented as fact. That it was "in the news" makes whatever is presented all the more believable.

That the news must be allowed to flow freely is why the inventors of modern democracy, both in France and America, made a special point of protecting the freedom of the press. For example, France felt that: (Italics added)

> " *'The free communication of thoughts and opinions is one of the most precious human rights: hence every citizen may speak, write, print with freedom, but shall be responsible for such abuses of this freedom as shall be determined by Law.'*
>
> "Freedom of speech, thus defined by Article 11 [above] of the 1789 Declaration of the Rights of Man and of the Citizen, has achieved universal scope worldwide. The article inspired the Universal Declaration of Human Rights adopted by the United Nations on 10 December 1948 (Article 19) and the European Convention on Human Rights adopted on 4 November 1950 (Article 10)." [46]

Information, including that which is untrue, must be allowed to flow unfettered. Thus we are not saying that falsehood in the news media should be made illegal—only that media truth ratings should be available to concerned citizens, so they know which sources they can trust.

This need not require evaluation of 100% of the news, which would be prohibitively expensive. A small random sample can accurately measure the level of truth within a small range, like plus or minus 3%, just as polls can measure how a population feels about an issue.

Once a workable approach to media truth ratings is introduced, a race to the top in the news industry will begin.

Let's return to the main strategy for this solution element: politician truth ratings. The truth of political arguments is not the only behavior that needs to be rated in order to establish the correct feedback loops. The overall corruption of politicians must also be rated. This is done with:

The Corruption Rating Solution Element

A **corruption rating** is an overall measure of how corrupt a politician is. Corruption includes falsehood, favoritism, coercion, abuse, criminal activity, the giving or accepting of bribes, knowledge that corruption is going on, and so on.

A major component of a politician's corruption ratings is past truth ratings. This would account for 40% or so of the rating. As a politician's truth ratings go up, his or her corruption rating would go down.

Corruption ratings would need to be done regularly, perhaps every two years. The running average of the last ten years or so would be a politician's rating. Corruption ratings would become as routine and cost about as much as a high level security check.

Truth ratings and corruption ratings are examples of **politician ratings**. They would be calculated in a similar manner by certified independent organizations. Both could cause the race to the top to become dominant. Because it measures total corruption, corruption ratings would play the stronger role. However truth ratings are easier and cheaper to perform, and thus would probably make a difference first.

Politician ratings need not affect all voters to make the critical difference—only the swing voters, who are normally just 10% to 30%. Fortunately it is this group who is most likely to be receptive to a tangible, sound reason to choose one politician over another.

The Analogy of Credit Ratings

Politician ratings are analogous to credit ratings. To demonstrate how important credit ratings have become in just one area, the corporate bond market, here is an excerpt from testimony presented to the US Senate on March 20, 2002, to the Committee on Governmental Affairs, chaired by Senator Joe Lieberman:[47] (Italics added)

"Simply put, a credit rating is an assessment of a company's credit worthiness or its likelihood of repaying its debt.

"John Moody, the founder of what is now Moody's Investors Service, is recognized for devising credit ratings in 1908 for public debt issues, mostly railroad bond issues. Moody's credit ratings, first published in 1909, met a need for *accurate, impartial, and independent information.*

"Now, almost a century later, an 'investment grade' credit rating has become an absolute necessity for any company that wants to tap the resources of the capital markets. The credit raters hold the key to capital and liquidity, the lifeblood of corporate America and of our capitalist economy. The rating affects a company's ability to borrow money; it affects whether a pension fund or a money market fund can invest in a company's bonds; and it affects stock price. *The difference between a good rating and a poor rating can be the difference between success and failure, prosperity and bad fortune.*"

In a similar manner, the difference between a good politician rating and a poor one would be the difference between success and failure for politicians, and prosperity and bad fortune for the public.

But even more interesting is the testimony went on to say:

"The government—through hundreds of laws and regulations—*requires corporate bonds to be rated* if they're to be considered appropriate investments for many institutional investors."

So too would the government require politicians to be rated if they were to be considered appropriate choices for many citizens. Credit ratings greatly lower the risk of financial loss. Corruption ratings would greatly lower the risk of corruption. If they proved as successful as credit ratings, they would lower it by somewhere around 99%, which would make sizeable cases of corruption about as frequent as Halley's Comet.

Presently corruption ratings are not required but corporate bond ratings are. This is one more example of how, over the centuries, the New Dominant Life Form has silently and relentlessly defined the rules of the game to be in its favor.

How Politician Ratings Work Dynamically

Like all deep structural change, politician ratings would cause important new feedback loops to become dominant. A diagram of these is shown on the next page. The main loop is **The Public Loves Those They Can Trust.** *This is probably the most important feedback loop in the entire solution, because if it works, the whole solution will probably work.*

Let's start at the top of the main loop, on the <u>use of ratings of politician's behavior</u> node. Suppose that node is activated because ratings have been implemented and are being regularly published for a few politicians. The ratings would at first be embarrassingly bad.

This would cause a rated politician to want to improve the quality of his or her behavior in order to get better ratings. This causes an increase in <u>virtuous behavior</u>, which would lead to <u>better truth and corruption ratings</u>. This would increase the <u>relative advantage of a politician in the eyes of the public</u>, because the public can now reliably tell whose arguments are more truthful and whose overall behavior is less corrupt, and thus who is a more trustworthy representative and more likely to get better results. This would increase <u>public support of the politician</u>, which would, in turn, increase their <u>election and reelection advantage</u>. The politician would know this happened. They would also know this benefited the people, so he or she would promote the <u>use of ratings of politician's behavior</u> so as to gain an even larger advantage and more benefits for the people. The loop then starts over.

Because politicians would now be competing to get better and better in the quality of their behavior, a race to the top among politicians would begin. This would cause the race to the bottom to collapse, because its supporters would switch to the race to the top.

The effect of ratings on the behavior of *Homo politico* would be astounding. That sub species would be singing "The public loves those they can trust, those they can trust," and other little ditties all the way to election day, and after that, to the next election day. *Homo citizenicos* everywhere would applaud, and join the chorus.

It is essential to understand the balancing loops that accompany the main loop. If problem solvers don't comprehend how the balancing loops work, they may be unable to design the most effective solution aspects, or they may have difficulty figuring out what went wrong if things go awry in implementation. *They may fail to understand what is limiting how far the race to the top can go, so they may be unable to make it go far enough.*

How the balancing loops work is too involved to cover in this brief chapter. For those curious about this, as well as the rest of the issues raised here, please see the manuscript for *A Model in Crisis*.

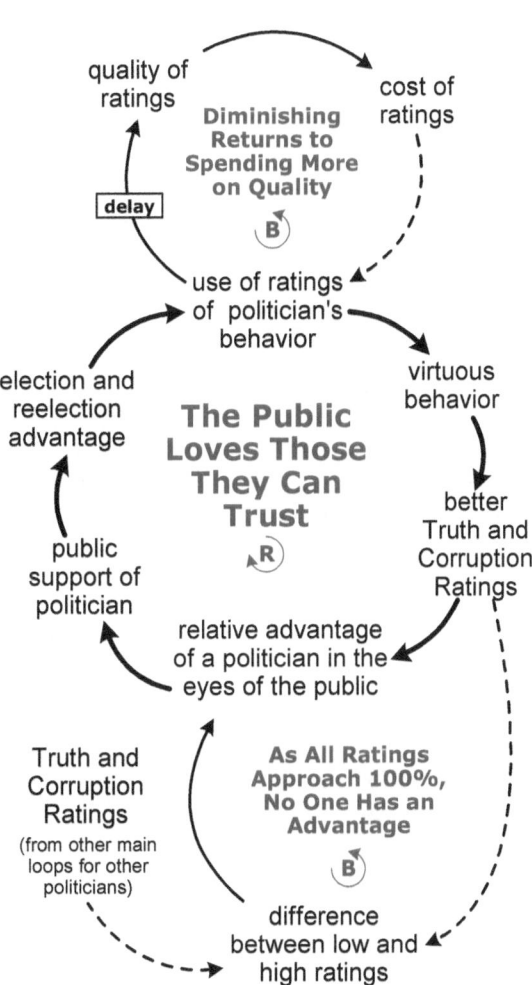

The Dynamic Structure of Politician Ratings

The three main loops of the politician ratings solution elements. This is deep, long overdue structural change to the human system. Like so many other fundamental feedback loop changes, such as voting and universal education, this change will automatically drive the system towards providing more for the greatest good of all.

Returning to our discussion, what if there is no way for truth and corruption raters to get the facts they need, because they are hidden behind a wall of secrecy? This is why we need:

The No Servant Secrets Solution Element

The objective of no servant secrets is to prevent servants, particularly politicians, governments, and corporations, from using secrecy to their own advantage.

This is accomplished by complete openness in all that a servant does. *No servant may keep competitive secrets of any type, either from their masters or other servants.* After all, if a servant is an entity created or employed by the hand of man to provide him with goods and services, why should a servant need to keep any form of competitive advantage secret, except to gain advantage over its master or other servants?

Competitive secrets are a form of non-sharing and hence a form of non-cooperation. When combined with the mutually exclusive goals that servants have of each maximizing something, such as profits, this leads to a *destructive competition* mindset. But what we want is *constructive competition*, where agents compete in a friendly, let's help each other manner. It appears that removing competitive secrets takes independent agents one step closer to cooperation. Therefore full and complete cooperation between servants and their masters, as well as between servants, requires no competitive secrets.

No servant secrets is short for *no competitive servant secrets*. It covers many areas. Some could be tackled soon. Others would take time. A few are counterintuitive and controversial, though less so as the analysis and solution strategy is more fully absorbed. Ultimately all would be dealt with, because a servant that keeps competitive secrets from its master has time and time again proven to be a danger to its master. The transition would probably take several generations.

No servant secrets is part of the Servant Realignment Package, which has eight solution elements. Together these serve to reengineer the modern corporation so that its interests no longer conflict with those of *Homo sapiens*. Because there are so many elements, a very flexible, as-needed approach can be taken.

No servant secrets is already spontaneously appearing in the form of freedom of information acts, sunshine laws, sites like OpenTheGovernment.org, the Federation of American Scientists project on government secrecy, and so forth. But these are a haphazard collection of ways to reduce servant secrecy.

Competitive secrecy needs to be reduced to zero in a comprehensive manner, which no servant secrets finally does.

One type of servant secret is government secrecy. A standard objection to eliminating government secrecy is the need for "national security." However this objection is really designed to benefit one country (and its military industrial complex) at the expense of others. Military secrecy is a form of competitive advantage. If countries truly want to cooperate instead of compete, then there is no need for military secrecy.

The standard rebuttal to this argument is that if I can't keep secrets and my competitor can, then they will gain an advantage over me. Rubbish. The same logic can be used to argue if I can't steal and my competitor can, they will gain an advantage. We have all seen that it is to society's benefit as a whole to outlaw theft. The same is true for secrecy. A country insisting on military secrecy is a country refusing to cooperate for the common good of all.

Because national security secrets increase the destructive competition mindset, they increase international conflict and/or preparation for it, which in turn increases the sales and profits of military goods and services. This benefits the military industrial complex, and hence the New Dominant Life Form. But it does not benefit Homo sapiens. In fact, international conflict or the diversion of national output to military purchases (the guns or butter choice) does just the opposite.

Servants include corporations. No servant secrets would mean the end of all competitive corporate secrecy. No longer could corporations ply politicians with secret favors and donations, or secretly influence political decision making. No longer could they secretly receive political favors. Because all this would now be out in the open, it would stop, because corporations are loathe to draw criticism from the people or the press.

Corporate secrecy includes trade secrets, which would no longer be allowed. The standard defense of trade secrets is they are necessary to provide an incentive for invention. Without trade secrets, a corporation could not make enough profit to pay for innovation.

This argument is fallacious. If corporations are servants and are truly working for the good of their masters, then the incentive to innovate should come from the desire to serve their masters the best they can, rather than to serve themselves as best they can. Trade secrets are really a form of selfishness.

Trade secrets are not necessary for scientists to innovate. Nor were they necessary for the long history of innovations that occurred up to modern times.

The real reason corporations want trade secrets is they are a form of competitive advantage. This greatly increases profits. But why should humans allow their servants to have any form of competitive advantage over other agents, which includes humans? There is no good rebuttal to that or the points raised above. Therefore trade secrets are not necessary and, because they are a form of secrecy that can be abused, they would not be permitted.

If any type of competitive advantage servant secrecy is allowed, then servants can use that as an excuse to hide all sorts of corruption from their masters. Thus no servant secrets means exactly that: no competitive servant secrets of any kind.

Certain forms of non-competitive advantage servant secrecy would be allowed, such as passwords. This is because passwords serve as identification and ownership identifiers, rather than as a form of competitive advantage. Other allowed types involve personal information, law enforcement, jury deliberations, and so on.

A special note: Several careful readers have suggested that the section on no servant secrets be removed because it makes it too easy for the opposition to find a spot to attack successfully. But without no servant secrets, there is no way to fully and accurately implement truth and corruption ratings. If servant secrets continue to be allowed, so much of the data needed for ratings will remain hidden behind a wall of secrecy that ratings will probably fail. Thus no servant secrets is a required prerequisite for creating the key new feedback loops necessary to eliminate the current dominance of the race to the bottom.

* * *

Let's assume that we have implemented the first five solution elements. These are freedom from falsehood, the Truth Test, truth ratings, corruption ratings, and no servant secrets. Would this be enough to raise the level of ability to detect political deception to a high enough level to solve the global environmental sustainability problem?

Not quite, because it lacks a measure of problem solving success. Lack of this has allowed many politicians (really corporate proxies) to more easily deceive the public with false priorities, and has dissipated problem solving effort.

The measure of problem solving success would be:

The Sustainability Index Solution Element

The top problem facing humanity today is the global environmental sustainability problem, because due to large social and ecological delays, it must be resolved proactively *now* to avoid catastrophe later. To trick the public and politicians into not solving this problem now, there is a tremendous fear, uncertainty, and doubt (FUD) campaign underway. This campaign has been so successful that millions of citizens, corporate managers, and politicians have been hoodwinked into thinking that the problem does not even exist, is not that bad, is too expensive to solve, lies too far in the future to worry about, or is so full of uncertainty solution is not required. Environmental sustainability has become such a low priority, especially in the US, that it is no longer a significant factor in elections or the national agenda. The corporate FUD campaign has worked all too well.

But it could be stopped in its tracks if citizens and politicians could look up and see, every day, a number that told them point blank how bad the problem really is and a graph showing where the trend is going. The sustainability index would provide exactly that. It would be an accurate, universally understandable measure of how well society is doing on solving the global environmental sustainability problem.

Instead of *fear* about the problem being too expensive to solve, there would now be fear about the cost of *not* solving the problem. This would really be concern, not fear, because now citizens would be facing a known, measured problem.

Instead of *uncertainty* about the status or magnitude of the problem, there would now be easily understandable numbers measuring how sustainable the planet is.

Finally, instead of *doubt* about the accuracy of data, there would now be a strong sense of trust that the Sustainability Index was as correct as is humanly possible. And, instead of *doubt* the problem needs solving now, there would be just the opposite: a strong national or global desire to solve the problem as soon as possible.

While no single measure of environmental sustainability is perfect, it is possible for a single number to accurately summarize how sustainable society is on a global basis. This single measure is called the sustainability index. It measures how much of the earth's carrying capacity is being used. If the index is over 100%, then it is unsustainable. Currently it is about 125%, as shown on the next page, though more recent data (2005) has increased this to 139%. [48]

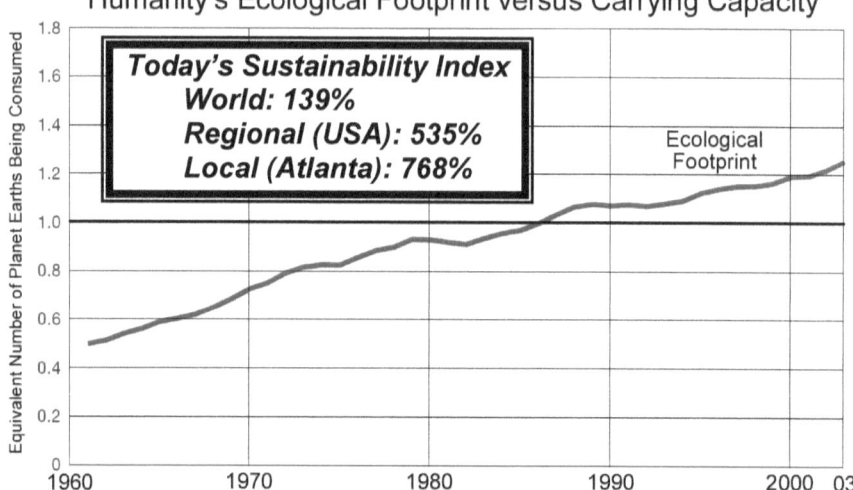

Here we have used the Ecological Footprint for the index, though any suitable index would do. The carrying capacity of the earth is approximated by the 1.0 horizontal line. This was crossed around 1985. It is not hard to visualize that if the footprint is extrapolated a few decades ahead, it will grow to such a high level of overshoot that catastrophic collapse is inevitable.

The index would include projected results (not shown). If society is doing nothing or too little to solve the problem, then people can immediately see that the projected Sustainability Index is still not good enough.

The sustainability index would be as widely published as stock market indexes. Eventually, once a suitable data collection system was in place, it would be updated just as frequently, in real time. Local, regional, and national indexes would also be published and compared. Together these would serve as a constant reminder of the true state of affairs, a sort of giant thermometer of the environmental health of civilization. (The local index shown above is estimated. The other two are from a different source than the graph.[49])

Further analysis may show that another index is also necessary or even better, such as a quality of life index. For most people this is what matters most, once their basic survival and security needs are met. A quality of life graph would probably show that sometime in the late 20th century it started going down a little, and is projected to go down a lot as the 21st century unfolds. Meanwhile, profits have been going up for the New Dominant Life Form. Showing these two curves on the same graph would have an enlightening effect, because it would become clear which life form was benefiting the most from the relationship.

How the Sustainability Index Works Dynamically

The purpose of the sustainability index is to provide an accurate, universally understandable measure of how well we are doing in solving the global environmental sustainability problem. Once the index is created, the **We Need to Be Sustainable** loop shown below will appear.

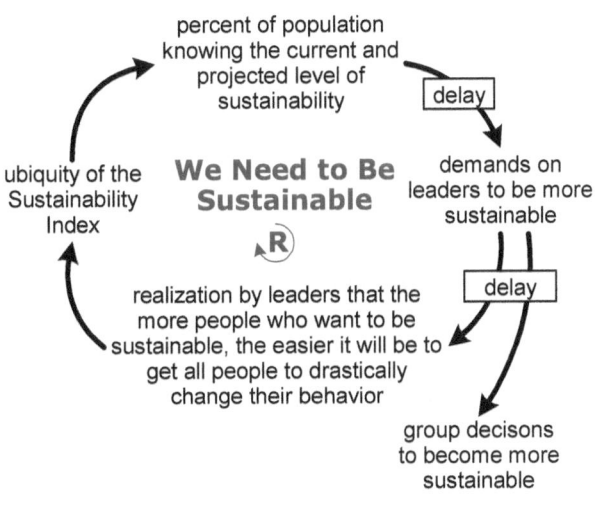

Actually many sustainability indexes or their equivalent already exist. Unfortunately they are not in the public's eye every day, mainly due to wrong priorities. Most are not sufficiently mature or updated frequently enough. If the wrong priorities of the race to the bottom can be changed to the right priorities of the race to the top, high quality sustainability indexes will start springing up faster than cornstalks in the springtime.

Starting at the left node, the loop works like this: When the index starts to be widely published, the ubiquity of the Sustainability Index goes up. This increases the percent of the population knowing the current and projected levels of sustainability. Due to a *delay* little will change at first, because it takes time for people to come to new conclusions. That is, it takes time for their sustainability memes (a meme is a mental belief) to grow in strength and number. But once those memes grow and reach a certain threshold of activation, people will increase their demands on leaders to be more sustainable.

Once again, little will change at first, because it also takes time for leaders to come to their own new conclusions. Their sustainability memes must grow in strength and number too. They must also grow to a high enough quantity and strength to overcome the competing memes emanating from the New Dominant Life Form.

But eventually, after a *delay*, this will happen, causing an increase in realization by leaders that the more people who want to be sustainable, the easier it will be to get all people to drastically change their behavior. One way to do that is to increase the ubiquity of the Sustainability Index, and the loop starts over again.

The loop also affects a node outside the loop. As <u>demands on leaders to be more sustainable</u> grows, so does <u>group decisions to become more sustainable</u>. This is the real benefit of creating the loop.

As the loop grows, more and more citizens and leaders will be thinking **We Need to Be Sustainable**. As the percentage of the population thinking this way becomes the majority and then a super majority, the desire to be sustainable will become an irresistible, unstoppable force that will lead to rapid solution of the problem. This will occur even if a large amount of self-sacrifice is necessary, because people will now see sustainability as the highest priority. They will see it this way because the alternative of not doing enough to solve it will be clearly shown by Sustainability Index projections as a certain road to disaster.

Summary and Conclusions

The six solution elements presented have been engineered to work closely together to change the general ability to detect political deception from low to high. These elements change the structure of the human system so that its new equilibrium is a state of high ability to detect deception. Once ability to detect deception goes high enough, the race to the bottom will collapse, causing the race to the top to become the dominant loop in politics. This in turn will lead to an intense global effort to solve the environmental sustainability problem.

Actually, these six solution elements are only part of the overall solution. Due to space limitations the solution presented here is incomplete. The full solution requires several dozen solution elements. There are also more high leverage points than the single one used here. We have presented only the first package and the most important high leverage point here. This is probably sufficient to get the ball rolling in the right direction, but not fast enough. Nor is it a permanent solution. For the reasons why and the other solution elements, please see the additional material at Thwink.org.

However, please note that this material is not that concerned with the exact solution. Instead, it focuses the bulk of its efforts on developing a *problem solving path* which, if taken, should quickly lead to an adequate solution. Our work emphasizes again and again that the solution presented is only a sample educational solution, and should not be interpreted as *the* solution. *This is because the fundamental reason for solution failure is the problem solving approach that most problem solvers have been using.* This is basically an ad hoc, common sense, event oriented approach. This works fine for everyday problems, but usually fails disastrously for difficult complex social system problems, such as the global environmental sustainability problem.

It is time for a thoughtful few words about that problem.

The political decision making process we use today was designed by the forces of evolutionary experimentation, one trial and error at a time. It is no more that a vast, ramshackle collection of historical precedent. Thus it is well designed to handle what it has encountered in the past. But it is ill prepared to handle problems which differ radically from those of the past, such as global environmental sustainability.

As a result, just when we need the political system to be working at its best, it is working at its worst. In most countries, highly partisan conflict frames legislative debate. In industrialized countries, behind the scenes the modern corporation and its allies control most of the key agents participating in that debate. This causes decisions to favor the interests of the New Dominant Life Form over the interests of *Homo sapiens.* Consequently what should be the political system's top priority, solving the global environmental sustainability problem, is barely on its radar.

It is time we threw off the backward looking forces of evolution as the chief designer of the political decision making process, and replaced it with the forward looking forces of engineering.

This may look hopelessly naive and impossible. Where do we start? How do we do it?

Those questions will remain unanswered as long as problem solvers continue using an ad hoc, common sense, event oriented approaches. But if they switch to the same stunningly successful approach that science adopted in the 17th century—rationality, through the use of a process that when correctly applied guarantees results—these questions could be answered.

The answers might be much like the six tightly coupled solution elements presented in this brief chapter. Out of millions of possibilities, these six were converged upon by the persistent application of the System Improvement Process and the continuous improvement of that process as it was applied. *It is only the output of a rigorous, highly refined engineering process like this that has any hope of solving a problem that has reached the very edge of a precipice.*

But as insightful as these answers may be, they are incomplete, because the Dueling Loops model is only half of what's needed to sufficiently understand the phenomenon of systemic change resistance. As the next chapter explains, the other half involves the most important contest our species has ever encountered: the battle for control of the biosphere.

Part Three

The Niche Succession Model and Sample Solution

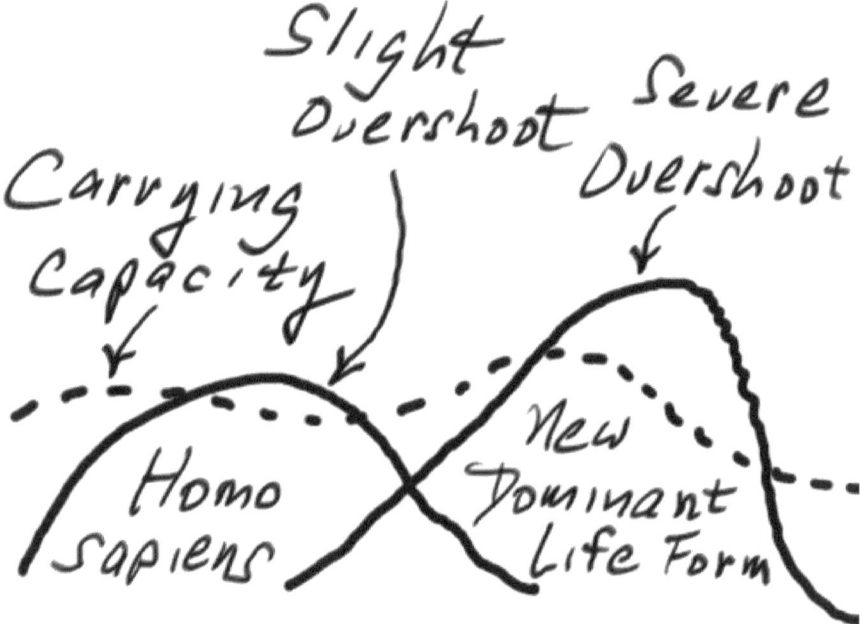

Chapter 9

The Battle for Niche Succession

L OOMING OVER THE ENDLESS DUEL OF THE POLITICAL POW-
ERPLACE STANDS THE BATTLE FOR NICHE SUCCESSION. It is a
winner-take-all clash between the two mightiest life forms on Earth: the modern corporation and *Homo sapiens*. The winner gains control of the biggest
niche on the planet: the biosphere. The loser has two choices: extinction or
adaptation to a lesser role, such as servant or slave to the winner.

In nature, epic battles like this have occurred billions of times. Every time
two species compete for control of the same ecological niche, another Battle
for Niche Succession runs its course, and evolution takes one more step forward.

In the human system, politics is one long series of battles for niches. Every election is a battle. The niche is the power of office, a coveted goal, because
once in office politician have immense control over the system. This allows
them to not only help themselves in their own next battle, but to help their
allies. This leads to the evolution of battle strategies that boggle the mind in
their cunning and complexity. The use of falsehood and favoritism is one of
these strategies. The use of truth is an opposing strategy.

Our hypothesis is these are *the* foundational strategies for politicians. This
hypothesis is expressed in The Dueling Loops of the Political Powerplace
model. But politicians are mere foot soldiers for the dominant life forms they
are fighting for. Thus to make the model more complete it needs another subsystem to handle the battle between the dominant life forms. This is what The
Battle for Niche Succession subsystem model does. *It is a model that, once
understood, will allow progressives to begin pushing on the highest leverage
point in the system in the long run: quality of political decisions.*

The full model of the political system contains two major subsystems:
The Dueling Loops of the Political Powerplace and The Battle of Niche Succession. The Dueling Loops model presented up to this point has been a simplified version. In this chapter we will briefly present the full Dueling Loops
subsystem and then The Battle for Niche Succession subsystem, which includes an additional high leverage point. The next chapter will then present an
example of how this high leverage point might be pushed on.

The full Dueling Loops subsystem is shown below. Four more feedback loops have been added to bring the model closer to the behavior we see in the real world. Here's how these loops work:

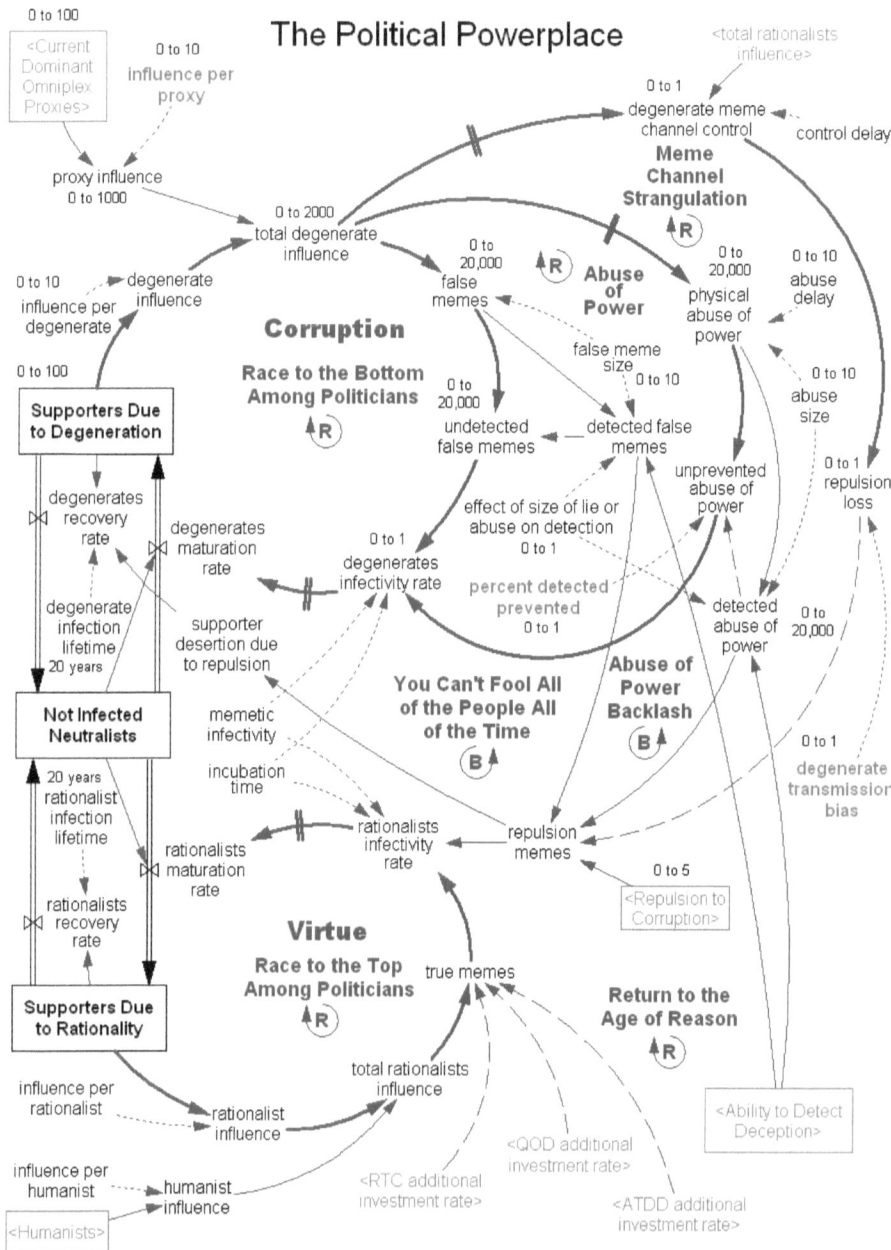

The Abuse of Power Loop

This is the loop that allows corrupt politicians to use <u>physical abuse of power</u> to win supporters, as opposed to the *mental* abuse of falsehood. The two main types of physical abuse of power are acts of favoritism and coercion.

Let's follow the loop around. <u>Total degenerate influence</u> times <u>abuse size</u> equals how much <u>physical abuse of power</u> is attempted. As the loop flows along this will become either detected or unprevented.

<u>Ability to Detect Deception</u> times <u>physical abuse of power</u> equals <u>detected abuse of power</u>. The rest of the attempted <u>physical abuse of power</u> goes to <u>unprevented abuse of power</u>. This then affects the <u>degenerates infectivity rate</u> in the same manner that <u>undetected false memes</u> do.

Abuse of power and false memes work almost the same. One difference is only some detected abuse of power can be prevented, but all detected false memes are prevented from infecting a new mind.

The Abuse of Power Backlash Loop

This works in an identical manner to the **You Can't Fool All of the People All of the Time** loop. People are repulsed by <u>detected abuse of power</u> as well as <u>detected false memes</u>.

The Meme Channel Strangulation Loop

Now things get interesting. The more a corrupt politician dominates the meme stream, the less their opposition can. Once they realize this, their best strategy is to flood all available meme channels with false memes. Examples of this are purchase and control of television, radio, and newspaper organizations, the creation of biased think tanks, the financial backing of prolific biased authors, the sponsorship of publicity events that lean their way, and the barrage of press releases that paint certain pictures. Once politicians are in power they can use the power of incumbency to dominate the meme stream even more.

All mass meme transmission channels in a society are controlled by someone. The model assumes that those controlled by uninfected supporters are *inactive* and don't matter. <u>Degenerate meme channel control</u> equals <u>total degenerate influence</u> divided by (<u>total degenerate influence</u> plus <u>total rationalists influence</u>). This varies from zero to 100%. <u>Degenerate meme channel control</u> times <u>degenerate transmission bias</u> equals <u>repulsion loss</u>. This also varies from zero to 100%.

For example if a corrupt political party had 40% of the population as supporters and a virtuous party had 20%, then *active* <u>degenerate meme channel</u>

control equals 40 / (40 + 20) = 66%. This means degenerates can decide what to transmit on 66% of a population's *active* media. (**Active media** is media that is actively trying to affect people's opinions.)

Things get even more interesting when we model how bias affects meme transmission. If the degenerates are unbiased then degenerate meme channel control doesn't matter. For example, if a news conglomerate owner is unbiased, then even though he controls a large chunk of the meme stream, he is not affecting it. But if he is a biased degenerate, then he can strangle the meme stream by transmitting more news that supports degenerates and less that supports virtuous viewpoints. This is common.

The Importance of High Quality Models

The detailed model descriptions may sound overly tedious and complicated, but they are merely the written form of the way our mental models work. Most people are not accustomed to thinking in terms of nodes and explicit relationships. System dynamics provides this discipline and allows our mental and physical models to grow to be complete and correct enough to solve problems that would be impossible to solve with mental models alone.

The purpose of physical models is to improve your mental models, which in turn improves the quality of your decisions. All conscious decisions are based on mental models of how the world behaves. *Therefore the better the model, the better the decisions.*

The model handles meme stream bias with the concept of repulsion memes. The more bias there is, the less truth that flows through the system about detected false memes and detected abuse of power. The less of this there is, the less people are repulsed by degenerates. This loss of memes that would normally cause repulsion is called repulsion loss. It equals degenerate meme channel control times degenerate transmission bias.

Repulsion loss is then used to calculate repulsion memes, which are what "push" supporters away from degenerates towards rationalists. Repulsion memes equals (1 - repulsion loss) times Repulsion to Corruption times (detected false memes plus detected abuse of power)

Repulsion to Corruption is a stock that varies from zero to 5. [50] It is a high leverage point with its own subsystem, which is not shown. Increasing Repulsion to Corruption increases repulsion memes, which increases the rationalists infectivity rate.

The model assumes that only corrupt politicians use transmission bias and that strangulation only affects repulsion. Strangulation could be modeled differently. It could also affect the <u>degenerates infectivity rate</u> and the amount of <u>true memes</u> transmitted. Degenerates can also control a higher percentage of the meme stream than their percentage of committed supporters would suggest. For simplicity we have not modeled these things, and leave them to a future iteration. The main result would probably be more advantage to the degenerates.

To summarize, meme channel strangulation has the effect of greatly reducing the effects of <u>Ability to Detect Deception</u> and <u>Repulsion to Corruption</u>. *Meme channel strangulation is the third inherent advantage of the race to the bottom.* The other two are false meme size and abuse of power size.

The Return to the Age of Reason Loops

After a thousand years of the Dark Ages, Europe returned to the Age of Reason in the second half of the 17th century. Also known as The Enlightenment, The Age of Reason emphasized the use of reason over dogma, and evidence over time honored assumptions that were too often false. According to wikipedia.com:

> *"The movement's leaders viewed themselves as a courageous, elite body of intellectuals who were leading the world toward progress, out of a long period of irrationality, superstition, and tyranny which began during a historical period they called the Dark Ages."*

It is time for a similar body of intellectuals to lead the world towards progress on the sustainability problem, out of a long period of irrationality (which has caused the masses to be too easily deceived), superstitious belief (in the miraculous power of economic and technological growth to solve all problems and lead to the highest good possible), and tyranny (by what we now know to be the New Dominant Life Form).

This can be done by strengthening what already exists. The human system already has feedback loops that cause problems of almost any type to eventually be solved. But the problem we seek to solve, the global environmental sustainability problem, has shown existing feedback loops to be too weak to respond correctly in time. Therefore to solve the problem there is little choice other than to very quickly strengthen these feedback loops, which we have chosen to collectively call the **Return to the Age of Reason** loops.

We have taken great care to incorporate a reasonable approximation of these feedback loops into the model. With proper engineering of how well they work, the race to the top can have another inherent advantage, one so strong it could rapidly lead to solution of the Why Such Strong Adoption Resistance problem. *The new advantage is the strengthened **Return to the**

Age of Reason loops. These are too complicated to present here. If the three high leverage points can all be raised, the net effect of these loops is to strengthen repulsion memes and total rationalists influence. This dramatically increases the true memes in the system, which causes the race to the top go dominant.

The Auto-Activation Chain

The three high leverage points are the stocks of Ability to Detect Deception, Repulsion to Corruption, and Quality of Decision Making. The first two may be seen on The Political Powerplace model. The third is on The Niche Succession model, which is presented later in this chapter.

These three high leverage points form an **auto-activation chain**. This occurs when activation of one reinforcing loop leads to activation of another, which leads to activation of still another reinforcing loop, and so on. This phenomenon is more commonly known as the *domino effect*. Here's how the chain works:

1. **Ability to Detect Deception** – We suspect this stock is the highest leverage point in the entire system. This means that the lowest amount of effort gives the highest amount of desired behavior. Therefore this is the first link in the auto-activation chain. It is **manually activated** by a specific problem solving project that causes the Ability to Detect Deception activation investment budget (part of the Ability to Detect Deception subsystem, which is not shown) to change to well above zero. The first link in an auto-activation chain must be activated manually. [51]

2. **Repulsion to Corruption** – Once the Ability to Detect Deception subsystem is activated, the level of the Ability to Detect Deception stock starts to grow. In the model it starts low, at 20% or 30%. After some years it reaches a medium level of 50% or 60%, which **automatically activates** the Repulsion to Corruption subsystem (also not shown). This causes the level of the Repulsion to Corruption stock to begin rising.

3. **Quality of Decision Making** – Ability to Detect Deception and Repulsion Corruption are both growing. Their product equals reaction to corruption. Once this passes the critical point in the Quality of Decision Making subsystem (also not shown), that subsystem is **automatically activated**, and its stock finally begins to rise.

As the auto-activation chain is activated, the population of degenerates starts to fall and the population of rationalists starts to rise. This continues until the chain is fully activated and the change resistance part of the sustainability problem is resolved. As a bonus, because the <u>Quality of Decision Making</u> is now quite high, solving the second part of the problem, proper coupling, will go much faster and better.

An auto-activation chain is identical to the managed phases of a large, well engineered construction project. Each subsequent phase builds on the one before it. When the project is done, the whole is now much greater that the sum of the parts, which gives the whole its new found efficiency. Generally this is several orders of magnitude greater than the parts working alone. This is how the auto-activation chain causes the three high leverage points to start working together in a hyper efficient manner, one that is strong enough to cause the **Return to the Age of Reason** to come to pass.

This completes the presentation of the full Political Powerplace subsystem. Next is the Niche Succession subsystem.

The Ecological Niche

First we must understand the concept of an ecological niche. In ecology, a **niche** is "a role claimed exclusively by a species through competition." The concept is so well established in the field of ecology that Stephen Jay Gould labeled it "the fundamental concept" of the discipline.

Ecology uses the niche concept "to address such questions as what determines the species diversity of a biological community, how similar organisms coexist in an area, how species divide up the resources of an environment, and how species within a community affect each other over time." [52]

The business world uses the concept to describe a "market niche." Firms battle it out for control of desirable market niches. Individuals use the concept, as in "she found her niche in the world." The concept has universal appeal, because it explains so much about agent competition in a finite environment.

A more complete definition is "The ecological **niche** of an organism is the position it fills in the environment, comprising the *conditions* under which it is found, the *resources* it utilizes, and the *time* it occurs there." [53]

Homo sapiens is one of millions of species that have successfully evolved to occupy a niche in today's world. Around 200,000 years ago, the species *Homo sapiens* (knowing man) diverged from the other species of *Homo: Homo habilis* (handy man) and *Homo erectus* (upright man), and the recently discovered *Homo floresiensis* (man of Flores, a remote island in Indonesia). These other species did not evolve as successfully, and became extinct, which is the fate of most species. [54]

Unlike all other species, *Homo sapiens* has developed the ability to highly modify the world around him, to in effect create a habitable niche nearly anywhere on the planet, and beyond into space. *Homo sapiens* has no genetic competition whatsoever for doing this.

But he does have memetic competition. As we have argued, *Homo sapiens* has already lost the battle to the New Dominant Life Form, which is the modern corporation and its allies. But *Homo sapiens* doesn't know this yet, causing him to continue to behave as a complacent employee and consumer, which is his new role. In other words, he has adapted and is now the indentured economic serf of the New Dominant Life Form.

What would be a good model of explanation for this forced adaptation? Such a model would need a centralizing, foundational concept. This is:

The Competitive Exclusion Principle

According to the **competitive exclusion principle**, when two life forms occupy the same niche, only one outcome is possible: One life form will drive out the other. If any of the other remains, it is only because its members have *adapted*, and are now living in a slightly different niche. Here's how the principle was discovered: (Italics added)

> "Georgyi Gause, the Russian microbiologist... interested in competition, discovered this principle. Gause inoculated a simple, finite culture with Paramecium, and... got logistic population growth. These Paramecium eat bacteria, and there is only so much food in a culture to support a certain number of Paramecium.
>
> "Then he put two [different] species of Paramecium in the same culture. He got lowered growth rates of both populations. *Even more interestingly, one species always drove the other to extinction.*
>
> "This led Gause to come forth with a famous 'principle' that would dominate ecological research for nearly the entire century: *Two species that use resources exactly the same way cannot coexist. One will drive the other to extinction."* [55]

This principle allows us to see what is really happening here. Two life forms, one genetic and one memetic, are battling for control of the biosphere. According to the competitive exclusion principle, the loser must adapt to a different niche or go extinct. There are no other choices.

It appears that *Homo sapiens* has chosen adaptation rather than extinction, so he is now subservient to the modern corporation and its allies. Depending on your point of view, his new niche is a powerless employee and consumer, a serf, or a slave. Perhaps it is all three. The major part of this transition is still in progress in the less industrialized areas of the world.

The data from one of Georgyi Gause's actual experiments is graphed on the right. The results tell a sobering story. [56]

Ecological Niche Succession

Once *Homo sapiens* ceded control of the biosphere to the New Dominant Life Form, an ecological niche succession event occurred. This has happened billions of times before in the genetic world, as one species overcame another in a struggle for

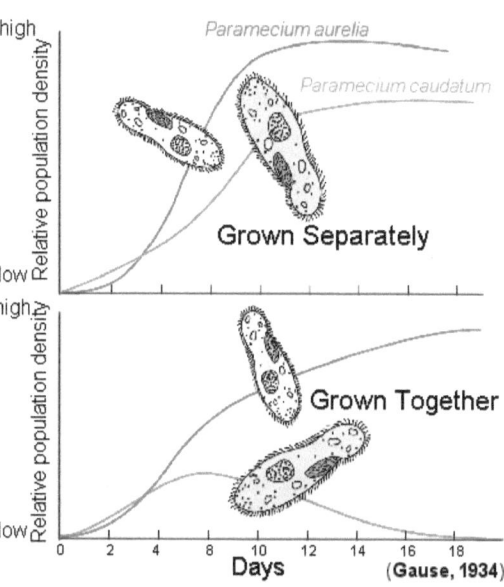

The lower graph shows the experimental results of competition between two species of Paramecium with similar requirements. Both did well for the first 3 days, but after that the species represented by the lower line was driven to extinction in 18 days, while the other species thrived.

survival in the same niche. Looking beyond the genetic world, it has probably happened trillions of times in the memetic world.

Niche succession occurs when successful competition from one life form drives another life form out of the same niche. This occurs due to superior strategies, superior physical abilities, or both. Sometimes luck is a factor.

The diagram below shows the idealized cyclic pattern of ecological niche succession as time goes by. The wavy horizontal dashed line is the population carrying capacity of the niche. The rising and falling curves are the populations of different life forms. The one with the most population (or influence, depending on how dominance is measured) is the dominant replicator. Except for during transition, there can be only one dominant replicator in a niche.

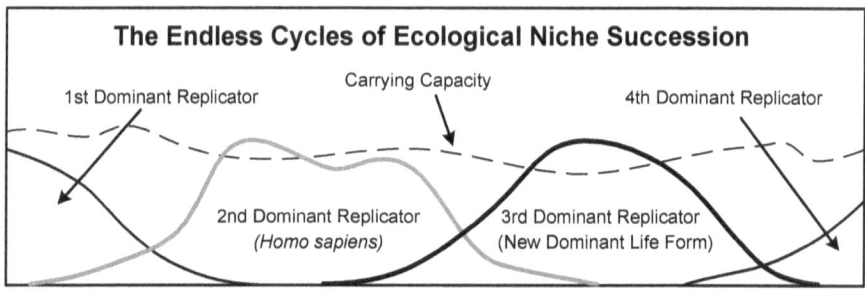

The Endless Cycles of Ecological Niche Succession

1st Dominant Replicator

Carrying Capacity

4th Dominant Replicator

2nd Dominant Replicator
(Homo sapiens)

3rd Dominant Replicator
(New Dominant Life Form)

On the left, the diagram starts with the 1st dominant replicator at the full population limit. At the same time, the population of the 2nd dominant replicator starts to grow from zero. As it grows, the population of the 1st one falls, falls some more, and goes extinct. The population of the 2nd dominant replicator grows to fill the niche and it enjoys exclusive control of the niche for awhile. It even exceeds the carrying capacity briefly, and then falls below it. Then another niche succession event starts, as the population of the 3rd dominant replicator starts to grow. The process is then repeated over and over indefinitely. It ends when the environment becomes incapable of supporting any form of life.

Substitute *Homo sapiens* for the 2nd dominant replicator and the New Dominant Life Form for the 3rd one, and you have the niche succession event in progress today.

The Three Choices

The theory of evolution says basically one thing: whoever adapts the best and the fastest wins. Adaptation is another word for evolution.

Ecology takes the theory of evolution one step further. It says that adaptation is to a particular *niche*.

Homo sapiens' three main choices for how this niche succession event plays out seem to be:

1. **Do nothing** – This leads to further submission to the role of economic serf/slave to the New Dominant Life Form, followed by eventual extinction. This would probably occur in two main steps. The first would be population collapse due to carrying capacity overshoot and extreme degradation of the biosphere. The second, which would occur even if the first was avoided or recovered from, would be due to eventual replacement of *Homo sapiens* by more efficient slaves, such as robots and genetically engineered and well trained animals.

2. **Adapt to another niche** – Instead of control of the biosphere, *Homo sapiens* is ideally suited for the coveted role of Top Slave to the New Dominant Life Form. If he plays this role well, becomes indispensable to his new master, learns how to say yes to everything, serves breakfast, lunch, and dinner to his master promptly, and thereby carves out a secure niche for himself, he may be able to spend the rest of his days in luxurious carefree servitude. This may be his best strategy, because at the moment people are far more capable than robots and animals. However human labor is very expensive.

3. **Adapt the modern corporation to a new niche** – This is a long shot, but it may not be too late to reengineer the modern corporation, so that it seeks a new niche. This would be as a trusted servant to *Homo sapiens* who puts the interests of his master before his own. This option takes full advantage of the concept of a niche, the principle of competitive exclusion, and *Homo sapiens'* greatest tool: reason.

The Ubiquitous Dueling Loops Pattern

Dueling Loops are a widespread pattern in social systems. Any two opposing groups of supporters can form this structure. There can be more than two, but two is the most common, because it is the majority versus the minority. Each group, the majority and the minority, has an incentive to stay united so as to either stay on top or try to get to the top. When staying on top is assured, the top group tends to splinter into two or more groups, because some in the top group want to climb even further. When a group has been on the bottom for awhile, they also tend to splinter due to the search for a better leader or strategy. Politics can thus be seen as the continuous change of the makeup and success of groups in the dueling loops structure.

Politics can also be seen as a continuous battle for niche succession of competing omniplexes. When two different life forms are locked in combat for control of a political system, the niche succession perspective provides the necessary explanatory power. For example, it explains why, as the battle between two life forms heightens, there is the tendency for *the end to justify the means*, and anything goes. This is to be expected, because the loser will die if it has no other niche to retreat to. (Actually the loser in politics no longer dies, ever since the invention of democratic elections. But many behave as if they will if they lose. What happens when they lose is they must adapt to the niche of being out of power.)

We have already examined one instance of the Dueling Loops pattern. This was the battle between corrupt and virtuous politicians. But there is another far more important Dueling Loops structure in the human system: The Battle for Niche Succession. There the current dominant omniplex and *Homo sapiens* are locked in an epic struggle to determine who will control the biosphere. The outcome of that struggle affects you, me, the entire population of the world, and all their descendants. As the ecological principle of competitive exclusion shows, the loser has to leave the game or become a slave to the winner. There are no other choices.

The Simplified Battle for Niche Succession Subsystem

The simplified model is shown below. First we will review it and then present the real subsystem.

The model's structure centers on two dueling loops. Instead of a race to the bottom versus a race to the top, two life forms compete for niche dominance. The niche they seek to dominate is the largest one imaginable: control of the entire biosphere.

The life forms are the New Dominant Life Form and *Homo sapiens*. Their supporters are corporate proxies and humanists. A **proxy** is someone who is heavily infected by an *omniplex* (defined below), causing the proxy to

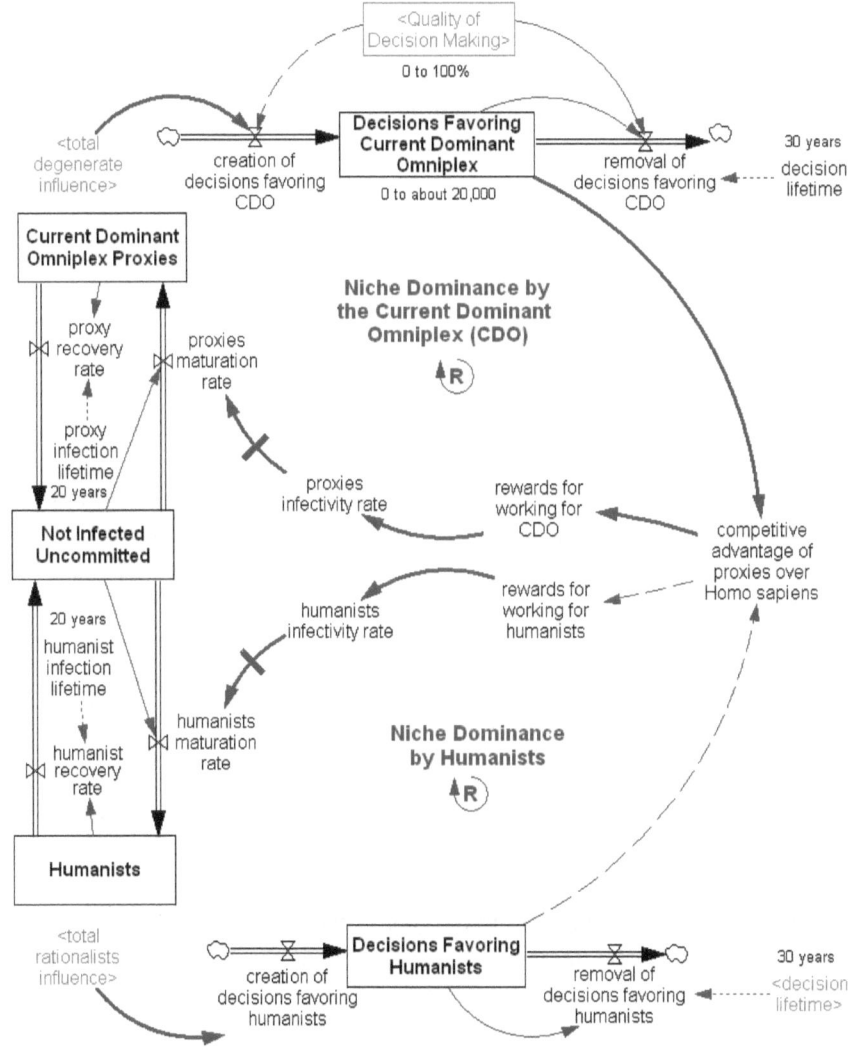

strongly support the omniplex's goals and directives in its everyday life. A proxy is thus an agent who repeatedly furthers the goals of an omniplex, whether it knows it or not. Proxies are also known as believers, the faithful, members, soldiers, and supporters.

Memes are described in depth on page 186187. A **meme** is a mental belief learned from others. A **memeplex** is a group or "complex" of memes. Examples are the ten amendments to the US Bill of Rights and the three fundamental steps of the evolutionary algorithm. The emergent properties of a memeplex are what make it important. For example, it is only through its three component memes of replication, mutation, and survival of the fittest that the memeplex of evolution gains its extraordinary explanative and predictive powers.

An **omniplex** is a memeplex that has copies in many minds and actively uses those minds as proxies to achieve its ends. An omniplex is short for omnipresent memeplex. Examples of omniplexes are ethnic cultures, religions, forms of government, political ideologies, and the modern corporation. Omniplexes are memetic life forms.

For the definition of a **humanist** we turn to Wikipedia, which says:

> "Humanism is a broad category of active ethical philosophies that affirm the dignity and worth of all people, based on the ability to determine right and wrong by appeal to universal human qualities—particularly rationalism. Humanism is a component of a variety of more specific philosophical systems, and is also incorporated into some religious schools of thought.
>
>   "Humanism entails a commitment to the search for truth and morality through human means in support of human interests. In focusing on the capacity for self-determination, Humanism rejects transcendental justifications, such as a dependence on the supernatural. Humanists endorse universal morality based on the commonality of human nature, suggesting that solutions to our social and cultural problems cannot be parochial [selfish or narrow minded]." [57]

Here is a quick summary of the model: Proxies throw their influence behind the degenerates in The Dueling Loops of the Political Powerplace subsystem. Humanists do the same for rationalists. Proxies and humanists each hope to swing the race to the bottom or top their way. There can be only one winner. Politicians will make more decisions favoring the winner, causing the winner to gain more and more supporters. This will rapidly lead to niche dominance.

Let's follow the top loop around, starting at Current Dominant Omniplex Proxies. This is used by the Political Powerplace subsystem to make the race to the bottom more dominant, by use of proxy influence. That dominance increases total degenerate influence, which increases the rate of creation of decisions favoring CDO, which increases Decisions Favoring Current Dominant Omniplex. This increases the competitive advantage of proxies over Homo sapiens, which is Decisions Favoring Current Dominant Omniplex divided by Decisions Favoring Humanists.

An increase in the competitive advantage of proxies over Homo sapiens increases the apparent rewards for working for CDO, which increases the proxies infectivity rate. After the time it takes for the infection to mature, the proxies maturation rate causes people to move from the Not Infected Uncommitted stock to the Current Dominant Omniplex Proxies stock. Then the loop starts all over again.

This is how an omniplex can employ people to increase its competitive advantage. The omniplex can be good or evil. Humanism is an omniplex.

The lower loop works the same way. The only notable difference is an increase in Decisions Favoring Humanists causes a decrease in the competitive advantage of proxies over Homo sapiens.

Both decision stocks have a variable that depletes them by obsolescence: removal of decisions favoring CDO or humanists. The decision lifetime is the same for both. It is 30 years.

That is the basic dueling loops structure. So far neither side has an inherent advantage, with one important exception: the influence of Quality of Decision Making. An increase in this causes a decrease in creation of decisions favoring CDO and an increase in removal of decisions favoring CDO. It could also affect Decisions Favoring Humanists, but for simplicity we have not modeled that, because it appears that would change the behavior of the model very little. This is because as long as an increase in Quality of Decision Making causes a decrease in Decisions Favoring Current Dominant Omniplex, the essence of the relationship has been established.

The Complete Battle for Niche Succession Subsystem

On the next page is the complete Battle for Niche Succession subsystem. The basic structure is the same. Two loops have been added. Here's how they work:

The Battle for Niche Succession

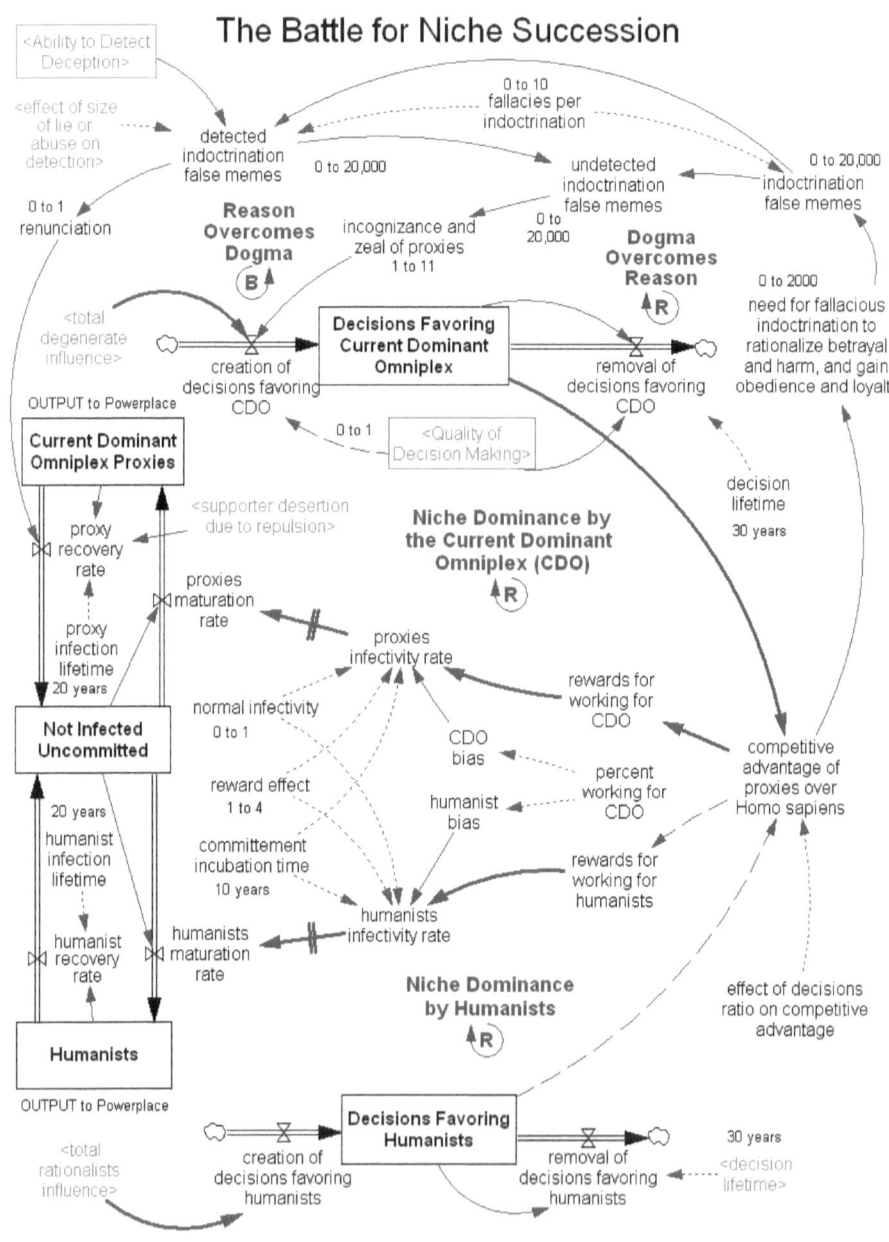

The Dogma Overcomes Reason *Reinforcing* Loop

At the risk of making the model too frighteningly realistic, we have added the well known phenomenon of the way dogma can overcome reason. According to Wikipedia: (Bolding added)

> "**Dogma** is belief or doctrine held by a religion or any kind of organization to be authoritative or beyond question. Many non-religious beliefs are often described as dogmas, for example in the fields of politics or philosophy, as well as within society itself. The term dogmatism carries the implication that people are upholding beliefs in an unthinking and conformist fashion. Dogmas are thought to be anathema to science and scientific analysis, and are strongly rejected by philosophies such as rationalism and skepticism. While in the context of religion the term is largely descriptive, outside of religion its current usage tends to carry a pejorative connotation—referring to concepts as being 'established' only according to a particular point of view, and thus one of doubtful foundation."

An unexpected thing happens once a person is infected by a dogmatic omniplex. A large part of the dogma has little to do with the benefits of believing in the dogma. *Instead, a huge chunk of the dogma is the fallacious indoctrination necessary to gain obedience and loyalty to the dogma.* Examples of this are intolerance towards non-believers ("you are with us or against us," which is a false dilemma), ostracizing or attacking those who disagree or question the dogma (such as by calling them unpatriotic or demented), the horrors of excommunication, continual demands towards a strict interpretation of the dogma rather than the use of one's own interpretation or common sense, and calls to keep one's faith in the dogma no matter what. This occurs not just in politics and religion, but to various degrees in science, art, ethnic cultures, and nearly everywhere. Few omniplexes are immune to this phenomenon. Why? Because once infected, this strategy causes believers to *stay* infected, *even if that is not in their best interests.* This can lead to a lifetime of involuntary servitude.

What if an omniplex is causing people harm? Many do. Examples are the fascist regime of Hitler, neo conservatism, the many religions that have waged countless wars, and the twisted ideologies of extremists such as terrorists and hate groups. To this can be added the modern corporation, when its obsessive pursuit of profit causes problems like child labor, sweatshop conditions, pollution, natural resource depletion, excessive inequality of wealth distribution, or war.

If an omniplex is causing people harm, it must engage in the <u>fallacious indoctrination necessary to rationalize betrayal and harm</u> (as shown on the model), or it will lose its adherents. This may seem difficult, but a clever dogma can do it easily. For example, the dogma just says, "There have been some problems here and there, but overall a rising economy lifts all boats. It's not a perfect world." Or it might say, "If we do not have a strong military, we will be subject to the whims and fancies of the rest of the world." And so on.

A widely practiced and very successful example of <u>fallacious indoctrination necessary to rationalize betrayal and harm</u> is the political philosophy of Leo Strauss (1899 to 1973). Here is how Straussian philosophy fallaciously justifies:

The need for continual war – "Because mankind is intrinsically wicked, he has to be governed. Such governance can only be established, however, when men are united – and they can only be united against other people." [58]

This explains why so may rulers want to keep their country in a perpetual state of war, or if war is not allowed, then a perpetual state of anxiety of some sort against a false enemy.

Lies are okay – "The essential truths about society and history should be held by an elite, and withheld from others who lack the fortitude to deal with truth. Society, Strauss thought, needs consoling lies." [59]

This is the infamous "noble lie" justification. Shadia Drury of the University of Calgary, author of *Leo Strauss and the American Right, 1999,* says "Perpetual deception of the citizens by those in power is critical (in Strauss's view) because they need to be led, and they need strong rulers to tell them what's good for them." [60]

The end justifies the means since there is no morality – In the above article Drury goes on the say that Strauss, like Plato, taught that within societies, "Some are fit to lead, and others to be led." But unlike Plato, who believed that leaders had to be people with such high moral standards that they could resist the temptations of power, Strauss thought that "those who are fit to rule are those who realize there is no morality and that there is only one natural right, the right of the superior to rule over the inferior."

The need for religion – Drury says that for Strauss, "Religion is the glue that holds society together." She adds that Irving Kristol, among other neoconservatives, has argued that separating church and state was the biggest mistake made by the founders of the U.S. republic. "Secular society in their view is the worst possible thing", she says, because it leads to individualism, liberalism and relativism, precisely those traits that might encourage dissent, which in turn could dangerously weaken society's ability to cope with external threats. Dissent would also cause a leader to lose power. "You want a crowd that you can manipulate like putty."

In other words, if people have already accepted one dogma, such as religion, they will far more easily accept another dogma and a dogmatic leader. While religion does have its beneficial aspects, it also has its dogmatic side.

The role of dogma is expressed in the model by one node for the need for dogma, and other nodes for the false memes necessary to transmit the dogma. Let's walk through the **Dogma Overcomes Reason** loop to see how this works.

If an omniplex is causing people harm, those supporting that omniplex have *betrayed* their fellow man. The greater the harm, the greater the betrayal. This becomes a strong reason for believers to abandon the dogma. To prevent this from happening and increase obedience and loyalty, as its competitive advantage grows the omniplex has a stronger and stronger <u>need for fallacious indoctrination to rationalize betrayal and harm, and gain obedience and loyalty</u>. This is the need for dogma node. The amount of this goes up as <u>competitive advantage of proxies over Homo sapiens</u> increases.

As the need for fallacious indoctrination increases, <u>indoctrination false memes</u> increase. Indoctrinations have a size. It is <u>fallacies per indoctrination</u>. That times the need for fallacious indoctrination equals the <u>indoctrination false memes</u>. Then <u>Ability to Detect Deception</u> times <u>indoctrination false memes</u> equals <u>detected indoctrination false memes</u>. <u>Indoctrination false memes</u> minus <u>detected indoctrination false memes</u> equals <u>undetected indoctrination false memes</u>. That increases the <u>incognizance and zeal of proxies</u>, which increases <u>creation of decisions favoring CDO</u>. This increases <u>Decisions Favoring Current Dominant Omniplex</u>, which increases the <u>competitive advantage of proxies over Homo sapiens</u>, which increases <u>need for fallacious indoctrination to rationalize betrayal and harm, and gain obedience and loyalty</u>, and the loop starts over again.

This has been a long story for such a minor loop, but that's how the **Dogma Overcomes Reason** loop works, both in the model and the real world.

The Reason Overcomes Dogma *Balancing* Loop

This loop serves to balance the **Dogma Overcomes Reason** reinforcing loop. That loop increased **Decisions Favoring Current Dominant Omniplex**. The **Reason Overcomes Dogma** loop decreases it.

Rather than trace this loop all around, let's start at <u>detected indoctrination false memes</u>. An increase in this causes an increase in <u>renunciation</u>. This increases the <u>proxy recovery rate</u>, which decreases the <u>Current Dominant Omniplex Proxies</u>. This in turn decreases <u>total degenerate influence</u>, which decreases <u>creation of decisions favoring CDO</u>, which of course decreases <u>Decisions Favoring Current Dominant Omniplex</u>.

This loop is the equivalent of the acts of renunciation we see in the real world, when dogma believers see the fallaciousness of their indoctrination at last, wake up, and go running towards the exit.

This loop has a crucial dependence on the <u>Ability to Detect Deception</u>. As this rises, so does <u>renunciation</u>.

There are no dogma loops at the bottom of The Battle for Niche Succession, as dogma by definition is fallacious. Humanists have no need for fallacious strategies, because once you fully understand the model (or intuitively grasp it, as has been the case for millions of progressives) and its high leverage points, the winning strategy (in the long run) is to stick to the truth.

The Dynamic Behavior of the Full Model

The behavior of the full model is about the same as the previous model. The main differences are the emphasis on the auto-activation chain and the presence of the Battle for Niche Succession subsystem.

Rather than review all the interesting simulation runs for the full model, we will examine only the run that represents solving the change resistance part of the problem. This requires that all three high leverage points become activated. This causes the model to respond very favorably, as shown on the next page.

As shown on the next page, the simulation run begins with the system in equilibrium. In the year 2000 the first link in the auto-activation chain is activated. Let's first discuss the starting conditions, which represent how the human system is now.

The run begins with the <u>Decisions Favoring Current Dominant Omniplex</u> high, and the <u>Decisions Favoring Humanists</u> very low. The ratio is about 20 to 1. This may seem absurd until you stop to consider the *comparative* advantages the modern corporation has accumulated over the last several centuries, while *Homo sapiens* has accumulated very few. A table comparing *The Comparative Competitive Advantage of Two Life Forms Due to Favorable Decisions* is presented in the next chapter. Once you have studied it you will probably agree that 20 to 1 is about right, and perhaps too conservative.

Run 6 - 140 Year View of the Battle for Niche Succession

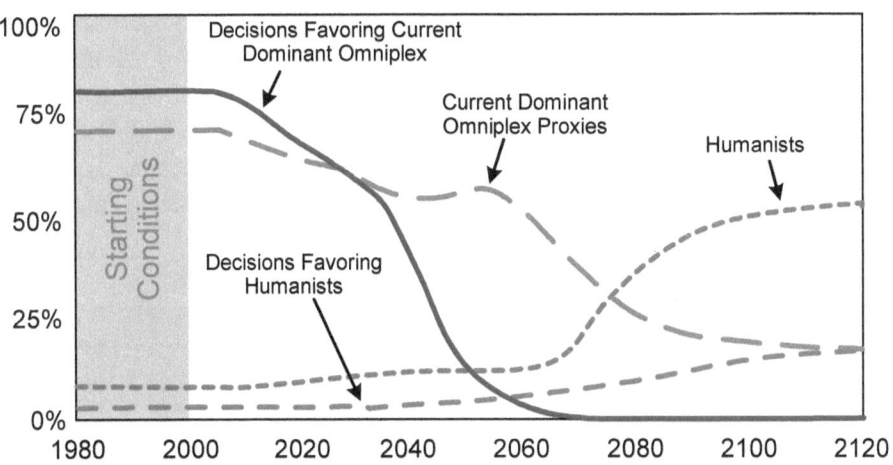

The run also begins with about 70% <u>Current Dominant Omniplex Proxies</u> and about 10% <u>Humanists</u>. This is the result of the lopsided competitive advantages listed in the table mentioned above. The result is the rewards for working for and furthering the interests of the corporate life form are much higher than for humanists, so corporate proxies outnumber humanists by about 70% to 10%. Poor *Homo sapiens*. He seems doomed to the role of corporate serf.

The model is not saying that the New Dominant Life Form has exactly 20 times as many decisions favoring it as *Homo sapiens* does. It's more like it has about an order of magnitude more, because this is a *relative model*, not an exact one. It shows relative differences rather than exact ones.

This run represents what may be humanity's best strategic path for solving the sustainability problem in time. *The input seems entirely possible to accomplish, because problem solvers are pushing on the three high leverage points in the optimum sequence, and so relatively little effort is required compared to the effort now being expended.* As the full auto-activation chain kicks in, the small force of pushing in the right places is leveraged into a much larger force that leads to solution of the change resistance problem. Tiny *Homo sapiens* has vanquished a much larger foe, not by the sword, but through use of mankind's greatest tool: reason.

Starting in 2000, as the three high leverage points are successively pushed, <u>Decisions Favoring Current Dominant Omniplex</u> takes a nosedive. After about 70 years it has fallen to near zero. Simultaneously <u>Decisions Favoring Humanists</u> rises, but not that much. But this doesn't matter, because on a relative basis humanists end up with over an order of magnitude of more decisions favoring them. As this change occurs, people notice. Due to delays in the system they don't react immediately, so the curve for the <u>Current Dominant Omniplex Proxies</u> drops a little later than the <u>Decisions Favoring Current Dominant Omniplex</u> curve. But eventually <u>Current Dominant Omniplex Proxies</u> falls to a low level and the <u>Humanists</u> curve rises to a high level. The run ends with the Battle for Niche Succession won by *Homo sapiens*.

Now let's venture over to the Political Powerplace subsystem. Things are pretty exciting there too, as seen in the graph below:

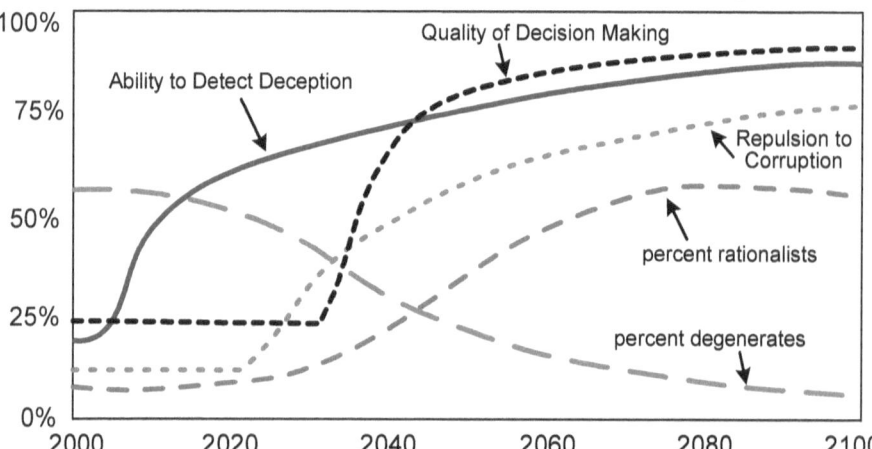

decisions favoring them. As this change occurs, people notice. Due to delays

Run 6 - 100 Year View of the Political Powerplace

We need to discuss this graph in some detail to understand how the auto-activation chain works. This is important, because *the solution is not good enough*. It takes over 40 years for <u>percent rationalists</u> to exceed <u>percent degenerates</u>. About the same is true for the other graph, where it took 50 years for <u>Decisions Favoring Current Dominant Omniplex</u> to fall below <u>Decisions Favoring Humanists</u>. Civilization does not have that long. There are large ecological thresholds that will be crossed before then, such as sizable melting of the Greenland ice sheet and the polar ice caps, massive deforestation and loss of photoplankton leading to lower ability to recycle CO_2, the release of CO_2 and methane from melting artic permafrost, and many more. Therefore we must figure out how the solution can be accelerated.

Here's how the auto-activation chain works:

Link 1 – The first step in the auto-activation chain is to dramatically increase investment in <u>Ability to Detect Deception</u>, starting in the year 2000. This causes <u>Ability to Detect Deception</u> to swing sharply upward after a delay of a few years. It grows swiftly until about 2010, then grows slower and finally approaches equilibrium at 2100.

This is a realistic curve. After the decision is made to increase <u>Ability to Detect Deception</u>, it takes years to develop new large scale social mechanisms to do that. They have to be conceived, built, tested, deployed, and then improved on an evolutionary basis as time goes by. Then for political deception to be actually detected by the masses and change their behavior, hundreds of millions or even billions of people have to start thinking in wholly new ways. This involves changing their minds on core beliefs, such as political ideology, attitudes towards the environment, and many habitual responses related to money and consumption.

This will take time and probably several generations. There is no other way. No one can reach into the human system and simply turn on a valve to make something grow to a high level, when that something has historically been low. Long delays like this must be anticipated and planned for, or possibly reduced by a dose of first-class social system engineering.

Notice that once <u>Ability to Detect Deception</u> starts growing a little, <u>Supporters Due to Degeneration</u> starts to fall and later <u>Supporters Due to Rationality</u> starts to rise. It takes <u>Supporters Due to Degeneration</u> over 20 years and <u>Supporters Due to Rationality</u> over 30 years to start changing significantly, even though <u>Ability to Detect Deception</u> jumped from 19% to about 45% in 10 years. Shouldn't the system respond much more quickly?

No, because the structure of the system makes it slow to respond. The main reason for the slow response would appear to be the 10 year incubation time for memetic infection. But an experiment changing this to one year shows it makes little difference. To make a notable difference you would have to go through the entire model and reduce many delays. This is very, very difficult to do in the real world. It may even be impossible.

Step 2 – The second step in the auto-activation chain occurs around 2020, when <u>Ability to Detect Deception</u> rises past the <u>true horrors revealed critical point</u>, which is 60%. It has taken it 20 years to grow from low to medium.

At this point the first automatic activation occurs. Once the critical point for more investment in <u>Repulsion to Corruption</u> is reached, more investment starts flowing automatically.

This is a critical point reaction. It represents the mass social realization that enough's enough, and we will stand for no more. Now that we can see what's happening (due to a 60% level of <u>Ability to Detect Deception</u>), we've got to rise up and throw all those corrupt politicians out, right now!

After the critical point reaction occurs around 2020, the level of <u>Repulsion to Corruption</u> turns upward and grows at a moderate pace for about 10 years. It then slows down to a leisurely rate of growth, reaching equilibrium in 2100. Except for a slower rate of growth and starting its growth later, it behaves the same as the <u>Ability to Detect Deception</u> curve. This is because they are both stocks, with identical subsystem structure. It is only the rate of investment in growth that varies. It is 15% for Ability to Detect Deception and 5% for Repulsion to Corruption.

Step 3 – The third link in the auto-activation chain occurs in 2032, when the product of <u>Ability to Detect Deception</u> and <u>Repulsion to Corruption</u> passes the <u>Homo sapiens sees the light at last critical point</u>, which equals 1.

The light *Homo sapiens* sees is not corruption. It is something far more important. What humanity sees is that it is the abysmally low level of <u>Quality of Decision Making</u> that has allowed the New Dominant Life Form to march in and take over. Unless humans can raise the level of <u>Quality of Decision Making</u>, they cannot turn the tables on their master. And if they cannot do it now, they probably never will. *This critical point reaction represents a massive united will to see the light and take full action at last.* Once it occurs, the entire problem is pretty much solved.

Once activated, <u>Quality of Decision Making</u> grows fastest of all, due to a 25% activation investment budget. It grows the fastest and tops out at the highest level of all the high leverage points, because in the long run it is the most important. It merely took the other two high leverage points to get it activated.

This completes inspection of the three auto-activation chain steps. There are many ways the steps could be designed. Different sequences, investment budget mixes, and critical points can be used. Investments can also be started simultaneously. There are also many more leverage points, which further study will no doubt find. This scenario is merely a simple, reasonable, first iteration.

Actually the manual activation in step one is an automatic reaction. *It is automatically activated when a small critical mass of people, on their own, personally take the effective action that ultimately leads to raising the level of* <u>Ability to Detect Deception</u> *high enough.* That spontaneous, unpredictable event is more easily represented in the model by the feature of manual activation. But we must remember that like so much else in the model, this feature is a simplification of reality.

I wonder who that small critical mass of people is going to be....

Summary and Conclusions

Adding the auto-activation chain, the Battle of Niche Succession subsystem, and the additional loops in the Political Powerplace has brought the model much closer to how the real world works. The purpose was to gain the further key insights necessary to solve the change resistance part of the environmental sustainability problem. The first insight is that the *root cause* of excessive model drift is low quality of political decision making.

An important finding is the sample solution takes too long. Using conservative, realistic parameter estimates, overcoming change resistance takes over 40 years. This is too long. Further analysis and experimentation will probably find several powerful innovative ways the solution can be accelerated. Will they be enough? No one really knows. But we do know that most societies that have gotten this close to environmental collapse have failed to change course in time.

The second insight is that quality of political decisions is the *long term highest leverage point.* Short term, the ability to detect political deception is the highest leverage point. The reason is there will be strong resistance from the New Dominant Life Form if any attempt is made to reduce decisions favoring it. Therefore large improvements in quality of decision making must probably come *after* an increase in ability to detect deception, which would make it more obvious to voters why a drastic increase in quality of decision making is needed.

However, there is the intriguing possibility that after a particularly bad case of corruption, the virtuous politicians voted in during the next election might be able to force a reform through so quickly that the New Dominant Life Form would be unable to resist. If the reform was based on deep structural change and pushed hard on the high leverage point of quality of political decisions, the reform might hold, and the New Dominant Life Form would be vanquished.

This is such a promising line of attack that the next chapter presents an example of one way problem solvers may be able to push on what is, in the long run, the highest high leverage point of them all.

Chapter 10

How to Raise the Quality of Political Decision Making

IMAGINE A POLITICAL SYSTEM WHERE EVERY POLITICIAN HAD THE INCENTIVE TO DO THE BEST JOB THEY COULD, not for themselves and special interests, but for the good of the system as a whole. Would that not lead to the fastest possible solution of the system's biggest problems?

Politicians are the people's elected problem solvers. Politicians, their staff, and the other politicians they work with are a problem solving organization working on one difficult problem after another. Because *the more difficult the problem the more mature the process used to solve it must be*, the best strategy is to use the most mature process possible. This will have the effect of maximizing the quality of political decisions, just as the many processes that corporations use serve to maximize their profits. If are you a politician and are serious about improving the quality of political decisions, then enlist the help of top corporate managers, because they are the best there is. Be sure to pick virtuous ones.

Presently political decision making quality is low, due to an immature process. This causes legislative decisions to be too easily controlled by corrupt politicians and special interests, notably proxies of the New Dominant Life Form who owe their allegiance to that life form instead of *Homo sapiens*. An immature process also causes the process to not adapt fast enough to changing times. This creeping obsolescence results in a growing inability to solve new types of problems, which leads to crisis management, bickering, and clever attempts to shift the blame for solution failure to others. A side effect of process immaturity is excessive partisanship, due to the focus of participants on personal or party gain instead of doing what is best for the whole.

All these problems would be greatly reduced if we could dramatically improve the political decision making process. This can be done with the decision ratings solution element. This has the complexity and power of double entry accounting and financial management, but because politicians manage something else entirely, it is totally different. However the principles of quantitative measurement of what you are managing, performance feedback, and continuous improvement apply equally well to political and business decisions making processes.

Decision Ratings

The objective of decision ratings is to improve the political decision making ability of governmental social control models, to the point where they can routinely solve difficult problems like sustainability. A **social control model** (defined on page 179) defines how a social unit runs itself, such as a family, corporation, or government.

The strategy is to create a race to the top among politicians to see who can accumulate the best decision ratings over their career. Under decision ratings, legislation undergoes a strictly monitored lifecycle. The lifecycle steps are objective, proposal, enactment, and outcome. For simplicity we will ignore solution evolution, solution management, obsolescence, etc.

Expert, non-partisan ratings are used to create powerful feedback loops over the course of a politician's career. The most important ratings occur early in a bill's lifecycle in the objective and proposal steps, when small improvements can have the greatest influence. This agrees with the fact that in legislative bodies the real work goes into drafting legislation, not in voting or managing solutions once they go out the door.

On the next page is a first iteration process map of the political decision making process using decision ratings. The best name for a process is not its department name, but a name expressing the beginning and ending states, or the input and output. For example, manufacturing is best called the procurement to shipment process. Sales is best called the prospect to order process. Thus this is **The Opportunity to Outcome Process**. [61]

The process steps with bolded borders are where the key politician or electorate work occurs. *If the decision points preceding these steps are of high quality and relevant data is available and also of high quality, then so is the work done in the bolded steps.*

Note the four reinforcing feedback loops identified by the Rs. These powerful forces drive the process toward higher and higher quality of decision making. *These loops are weak or nonexistent in the present process.*

Here's how Decision Ratings handles the lifecycle steps of a bill:

1. The Objective Step – Decision Ratings uses a *hierarchy of objectives*. At the top sits a nation's standing goals. These are enshrined in its constitution or a similar document. At the bottom are all the bills currently in force. Between the top and bottom is an implied but unwritten set of layers of intermediate objectives. Those doing the ratings work with legislatures to develop a published system for keeping track of the hierarchy. Eventually certain bills will probably be created to authoritatively define portions of the hierarchy.

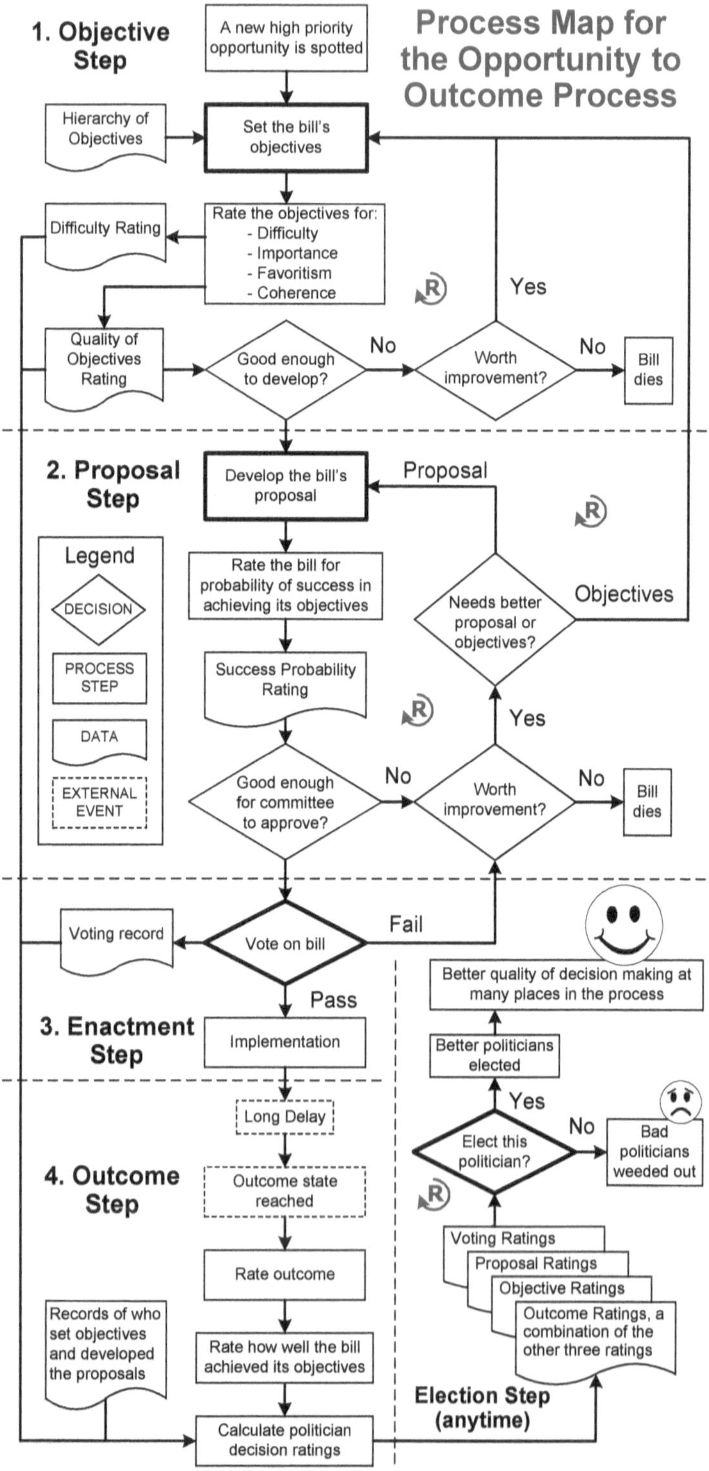

Process Map for the Opportunity to Outcome Process

1. Objective Step

A new high priority opportunity is spotted

Hierarchy of Objectives

Set the bill's objectives

Rate the objectives for:
- Difficulty
- Importance
- Favoritism
- Coherence

Difficulty Rating

Quality of Objectives Rating

Good enough to develop? → No → Worth improvement? → No → Bill dies

Yes

2. Proposal Step

Develop the bill's proposal

Proposal

Legend

DECISION

PROCESS STEP

DATA

EXTERNAL EVENT

Rate the bill for probability of success in achieving its objectives

Success Probability Rating

Needs better proposal or objectives?

Objectives

Good enough for committee to approve? → No → Worth improvement? → No → Bill dies

Yes

Voting record

Vote on bill — Fail

Pass

Better quality of decision making at many places in the process

3. Enactment Step

Implementation

Better politicians elected

Yes

4. Outcome Step

Long Delay

Outcome state reached

Elect this politician? → No → Bad politicians weeded out

Rate outcome

Voting Ratings
Proposal Ratings
Objective Ratings
Outcome Ratings, a combination of the other three ratings

Records of who set objectives and developed the proposals

Rate how well the bill achieved its objectives

Calculate politician decision ratings

Election Step (anytime)

When a bill is first created its objectives are set. These are then rated for four things: difficulty, importance, favoritism, and coherence. The last is how well the objectives support the existing hierarchy of objectives. The last three are then weighted to create an overall rating of *quality of objectives*. If any favoritism or irrelevancy exists, this will cause a low quality of objectives rating, because that would mean the bill's objectives clash with the hierarchy.

It will not be long before committees set a high quality bar, such as 90%, that the objective rating of all new bills must pass to be developed by committees into full proposals. Who created the objectives is recorded for later use.

2. The Proposal Step – After a bill is fully developed it becomes a proposal. It is then submitted to the raters who rate it on how likely it is to achieve its stated objectives, which is called its *success probability rating*. This is a type of **predictive rating**, as opposed to a rating based on how things turn out, which would be an **outcome rating**.

At first the raters must study the lifecycles of lots of past bills, calibrate their predictive process, and make educated guesses. With experience and specialization they will get better and better. The raters will themselves be rated by an independent body for how well their past ratings correlate with outcomes, which will allow a confidence level for a rater's ratings. Multiple rating organizations will specialize in different types of legislation and compete to see who can get the highest confidence levels, because that's who politicians are going to want to rate their bills. Proposals are also rated on favoritism.

Again, it will not be long before congressional bodies insist that a proposal must have at least an 80% or so probability of success and no more than a 5% or so favoritism rating before it may be brought to the floor for final debate and voting. [62] If a bill passes it moves to the next step.

Under these conditions we are going to see the instant disappearance of sneaky midnight earmarks, late amendments, and all the trickery that pops out of the sky when bills come out of committee. This is because if any change is made the proposal must be rated again. This takes days to weeks at a minimum, costs a considerable amount of money, and any favoritism or poor quality of decision making that has crept in will hurt the bill's ratings. If the probability of success falls too low or the favoritism rating rises too high then alternative bills will take its place or it will not be allowed on the floor.

The raters record who the authors are for each bill. The simplest way to do this is to see who is on the committee that created it. Better ways will evolve to reflect who did the real work and made or suggested the key decisions.

3. The Enactment Step – If a proposal passes, the raters record who voted for and against it.

4. The Outcome Step – Finally, years later, the raters measure the bottom line: how well a bill achieved its objectives. This is done for all enacted bills.

The results are then correlated with enactment votes to see who has the better record on voting for bills that better achieved their objectives. The correlation is then adjusted for the difficulty of the objectives. This gives the **voting rating** for each politician. The same thing is done for outcomes versus the records of who authored each proposal, plus adjustment for difficulty, which gives the **proposal rating** for each politician. Finally, the same thing is done for outcomes versus quality of objectives, plus adjustment for difficulty, which gives the **objective rating** for each politician.

Objective ratings are the most important, because they represent the strategies and priorities behind a politician's work. Next in importance are the proposal ratings. They represent the quality of the bulk of that work. The public will know this and weight the three ratings accordingly, probably around 50% for objective ratings, 30% for proposal ratings, and 20% for voting ratings. The three ratings might then be combined into a single *outcome rating*.

If Decision Ratings are implemented at the local, state, national, and ideally the international level, all politicians will have lifetime ratings. Voters will look long and hard at a candidate's ratings history as they make their choices. *They will probably consider ratings more than any other factor, because now they have an objective, reliable, understandable, comparable measurement of what they have always wanted to know: How well is a candidate probably going to do in the future to help achieve my society's objectives?* The result will be a race to the top among politicians to see who can accumulate the best decision ratings over their career.

One of the first things they will do is to say:

Goodbye to the Tremendous Competitive Advantage of Corporations

At the very top of the hierarchy of objectives is optimizing the system for the greatest good of all people. Once they have this goal firmly in mind, politicians will begin to see that the New Dominant Life Form has been pursing an entirely different goal: optimizing the system for the greatest good of corporations. The corporate life form has accomplished this by relentlessly changing the system to favor themselves over people. This has been done so cleverly and in such small, imperceptible increments that few citizens have noticed. But when you pause to examine the outcome, the findings are shocking, as the table below shows.

Only in the first attribute does *Homo sapiens* have the advantage. In the second attribute they are equal. In all the rest the modern corporation has the overwhelming advantage.

Galloping galoshes! Decision by legal decision the modern corporation has built up an astronomical lead over *Homo sapiens*. These are huge, order of magnitude advantages. There is little question who is going to win the battle for niche dominance unless things change. Furthermore, because corporations march to the beat of a different drummer (maximization of profit instead of maximization of quality of life), they have been aggressively using these advantages to their own benefit, with only enough regard for their opponent to keep him alive so that he may perform his role of incognizant slave.

The Competitive Advantage of Two Life Forms Notice how only in the first attribute does *Homo sapiens* have the advantage.		
Attribute	**The Modern Corporation**	***Homo sapiens***
Can physically manipulate its surroundings	No	Yes
Is legally considered a person	Yes	Yes
Maximum life span	Infinite	About 120 years
Can be in many places at the same time	Yes	No
Can own slaves like itself	Yes	No
Speed of procreation	Hours	Nine months
Can cut itself up into little pieces, each of which can become a new life form	Yes	No
Can hibernate indefinitely in hard times	Yes	No
Body size limit	Unlimited	About 8 feet high
Brain size limit	Unlimited	About 1,500 grams
Owners have limited liability	Yes	No, since no owners
Has international organization with high efficiency of decision making and full power of enforcement of decisions for its life form type	Yes, the World Trade Organization	No, the United Nations
Primary energy input	**Money** via sales	**Food**
Requires a physical form for its primary energy	No	Yes
Can transmit its primary energy instantaneously over great distances	Yes	No
Can store its primary energy indefinitely	Yes	No
Can store infinite amounts of its primary energy at no cost	Yes	No
Financial impact of storing its primary energy	Makes a profit by charging interest	Must pay storage costs

Now we can see why it is so crucial to improve the quality of political decision making. If it is low, then it is too easy to dupe politicians into decisions that favor corporations over people. If it is high, then they can no longer be so easily fooled.

But it may be too late to reverse the tremendous competitive advantage corporations have over people. We are now utterly dependent on them. They control our lives. If we attempt to turn the tables and put the corporate life form in its proper place, as a humble and loyal servant to *Homo sapiens*, what is probably going to happen? Like all tyrannical masters, it will resist. Strongly. How the battle for supremacy plays out is anyone's guess, but the odds do not favor *Homo sapiens*, because most revolutions fail, and are brutally suppressed.

There are, however, a few enlightened corporate managers who are humanists first and corporate proxies second. Perhaps they will be the Trojan horse that sees the merit of symbiosis instead of continued parasitic exploitation. If we are lucky they will lead the way from the inside, and usher in improvements like corruption ratings, no servant secrets, and decision ratings, which would begin to turn their master into the intelligent, mindful servant it was originally created to be.

How Decision Ratings Work Dynamically.

Once decision ratings are introduced elections will become non-events. They will be as exciting as watching paint dry and as predictable as your favorite cornbread recipe. The results will almost always be a foregone conclusion, except for first timers and very close ratings, due to the driving force of the published ratings. Voters will choose the best candidates fairly rationally, which implies what they are doing today. And they will do it at low cost, because there will no longer be an advantage to spending huge amounts of money and effort on painting the grand illusion that politician A is better than B, because of a hundred and one fallacious reasons. That money and energy is better spent elsewhere in the system.

Decision ratings have a surprisingly simple dynamic structure, as shown on the next page. The main loop is similar to the one for politician ratings on page 105. For simplicity the balancing loops are omitted.

Let's walk the loop, starting at <u>use of decision ratings to make decisions</u>. This node is first activated when decision ratings are first introduced in a government. The ratings would at first be very low. Use of the decision ratings process would improve <u>quality of decisions</u>. As this went up it would lead to <u>better predictive ratings</u> in the short term. In the long term, after a *delay* it would lead to

improved <u>quality of actual outcomes</u>. This would cause <u>better outcome ratings</u>.

The predictive and outcome ratings would be widely published. If a politician's ratings were better than their opponent's then that politician would tout them to their constituency. This would increase the <u>relative advantage of a politician in the eyes of the public</u>, because the public can now reliably tell whose work is more valuable. This would increase <u>public support of the politician</u>, which would in turn increase their <u>election and reelection advantage</u>. Politicians would know this has happened, giving them the incentive to promote the <u>use of decision ratings to make decisions</u> all the more. The loop then starts all over again.

Because politicians would now be competing to see who can get the best lifetime ratings, a race to the top would begin. *And it would never stop, because the process is self-improving.*

As the loops grow, politicians in other governments will notice the <u>election and reelection advantage</u> their fellow politicians are gaining, as well as the superior <u>quality of decisions</u> other cities, states, or countries are making. They will then spontaneously begin the <u>use of decision ratings to make decisions</u> in their own political systems. In this manner loop growth would cause decision ratings to spread across the human system faster than you can say "Follow the money," which would now be obsolete, because the new slogan for investigative reporting would be "Follow the ratings."

If this structure can be established then the sustainability problem and other difficult social problems will be solved, because the loops are self-improving. Once decision ratings start, the most important decisions in the loops will be those that improve the decision making process itself. This is because the most important step in any non-trivial process is continuous process improvement. This is such a fundamental principle that anything intelligent that evolves (including life forms and social systems) can be seen as a self-improving, self-managing process. Every time the evolutionary algorithm produces another mutation that improves the entity's competitive advantage, the process has improved.

Notice how thinking in loops lies at the very heart of how to radically improve complex social systems. Unless progressive activists become as good at this as they are at breathing, solving difficult social problems will remain as elusive as ever.

Summary and Conclusions

This brief sketch should explain how creating the right feedback loops can dramatically improve the quality of group decision making at all levels of politics. *The system will now have automatic accountability,* if voters use the ratings as I suspect they will. Imagine what the beneficial effects might be. And imagine what problem would already be solved if decision ratings already existed.

Decision ratings would cause a sea change in the way bills are developed. High ratings would require sound analysis of the causes of a problem, deep understanding of how people and systems behave, a thorough look at all reasonable alternatives, lots of synthesis to create new ideas, a method of picking the best solution path, and techniques to prove that all this is correct and not just highly plausible. Undue personal bias would not be allowed. This of course is exactly how successful corporations have worked for a long time.

If you are a politician and your government is making less than excellent decisions, then the most important item on your agenda should be to help create something like decision ratings. Or this short sketch may give you even better ideas. *Start simple.* For example, start with only the most important bills, only one legislative body, and only a few key objectives. Or consider adding this amendment to your constitution:

Using the principles of system dynamics, congress shall install a formal opportunity to outcome process on itself that drives congress toward optimizing the human system for the common good of all and their descendents.

Once deep structural changes to the system are made, then, and only then, will democracy have the foundation it needs to achieve what has never been possible. This amendment will quickly be seen as the most important one of all, because it maximizes the chance of achieving all the others.

Notice where the last three words in the amendment will then lead.

Part Four

How Can We Apply This New Knowledge?

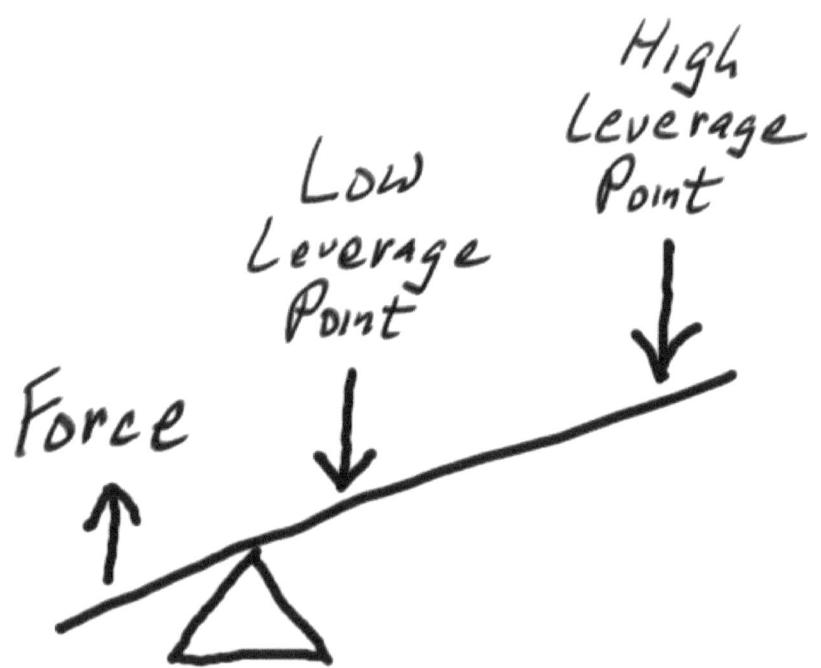

Chapter 11

The Assault on Reason Examined

IT IS TIME TO MOVE FROM THEORY TO APPLICATION. The dueling loops model allows us to see that most difficult progressive problems are a side effect of a dominant race to the bottom. Let's apply this principle to a specific problem: the behavior of the George W. Bush administration in the United States.

Al Gore, in *The Assault on Reason,* published in May of 2007, presents a penetrating look at this problem. Going far beyond the cursory issues where most writers dwell, he dives right to the core of the problem with the book's title. This is the central premise of his book: that reason itself is under assault and is losing.

The book begins by framing the problem this way: (Bolding added)

"Why do reason, logic, and **truth** seem to play a sharply diminished role in the way America now makes important decisions?

"The persistent and sustained reliance on **falsehoods** as the basis of policy, even in the face of massive and well-understood evidence to the contrary, seems to many Americans to have reached levels that were previously unimaginable.

"A large and growing number of Americans are asking out loud: 'What has happened to our country?' More and more people are trying to figure out what has gone wrong in our **democracy**, and how we can **fix it**." (p1)

Right away we see the use of "truth" has declined, the use of "falsehoods" is the dominant mode, and that "democracy" is broken and needs to be fixed. This agrees 100% with our analysis. If the use of truth is low, that indicates a weak race to the top loop. If the use of falsehood is high, that indicates the race to the bottom is dominant. This shows the model of democracy is in the Model Crises step of the Kuhn Cycle, as described on page 184. The model can be fixed by improving it so that it can resist whatever it is that has so badly broken the present model.

This is an impressive start. But how far can Gore and others like him go, without a formal model behind their analysis? As we shall see, they can only go so far before lack of a sound model leads to an incorrect diagnosis, which in turn leads to an incorrect solution. But this should not detract from the powerful message in the book: that reason is under assault, and until it is restored, democracy will be unable to deliver what its inventors intended.

The Diagnosis

The Assault on Reason demonstrates how, when it comes to solving incredibly difficult problems, the importance of asking the right questions cannot be overstated. It's what separates those who can solve such problems from those who cannot. For example, once doctors have gone through training, they can now ask the right questions, until they have determined WHY a patient's symptoms are present.

This is the all important diagnostic step. Once the patient has been correctly diagnosed, the next step, treatment, is usually relatively straightforward. But without a proper diagnosis, the doctor can only guess at how to cure the patient.

Thus Al Gore begins precisely where he should: with the penetrating, persistent diagnostic questions that far too few are asking. His mission in *The Assault on Reason* is to find out WHY the political process in America is so gravely ill and then how to cure the patient. To begin the diagnosis, he asks "Why do reason, logic and truth seem to play a sharply diminished role in the way America now makes important decisions?" Then he puts it another way:

> "We have a Congress. We have an independent judiciary. We have checks and balances. We are a nation of laws. We have free speech. We have a free press. Have they all failed us? Why has America's public discourse become less focused and clear, less reasoned?" (p2)

Why indeed? Unless we can find the answer, democracy is doomed. Gore's answer, his diagnosis, begins by observing that:

> "Our Founders' faith in the viability of representative democracy rested on their trust in the wisdom of a well-informed citizenry, their ingenious design for checks and balances, and their belief that the rule of reason is the natural sovereign of a free people. The Founders took great care to protect the openness of the marketplace of ideas so that knowledge could flow freely." (p5)

This leads to what Mark Twain would call the "nub" of the diagnosis:

"In practice, what television's dominance has come to mean is that the inherent value of political propositions put forward by candidates is now largely irrelevant compared with the image-based ad campaigns they use to shape the perceptions of voters. That is why campaign finance reform, however well drafted, often misses the main point: so long as the dominant means of engaging in political dialogue is through purchasing expensive television advertising, money will continue in one way or another to dominate American politics." (p8) [63]

No Solutions Here

This chapter is a short, timely example of the application of the two tools presented in this book: a process that fits the problem and simulation modeling. Thus moving from theory to application does not mean that we are about to solve the sustainability problem or any other problem. *It only means that we are trying to show how to better go about solving them.*

So there we have it: a diagnosis of WHY the system is broken. Money is dominating politics, because "expensive television advertising" is now what swings elections. Media ownership concentration and bias is also involved. Together, let's call these the **media factor**.

The Diagnosis Examined

I hesitate to say this, but I suspect this diagnosis is incomplete. Further WHY questions need to be asked: WHY is there a persistent tendency for money to dominate politics? WHY has this been the case long before television even existed? WHY does campaign finance reform usually fail? And then there's a much deeper question: WHY do politicians prefer to tell falsehoods rather than the truth to win elections?

Another "test the diagnosis" question would be WHY did the assault on reason jump so dramatically during the George W. Bush administration, starting in 2001? It can't be because of the media factor, because this was available to previous presidents to abuse, and the previous administration, for example, did not. In fact, no recent president has waged an assault on reason of the magnitude and ferocity the George W. Bush administration has. Furthermore, no significant changes in the media factor occurred around 2001. The proper diagnosis must therefore be something else.

The media factor is indeed *a* factor. But it is not *the* factor.

Like most, Gore has stopped at the first plausible, intuitively attractive diagnosis. This one has great appeal because the facts are true. Money IS dominating politics. TV ads DO swing elections. Media ownership IS too concentrated. But it does not follow that if we can prevent money from swinging elections via TV ads that the problem is solved. The money will simply be used to accomplish the same end in other ways. Political TV ad reform will fail, just as campaign finance and lobbying reform have repeatedly failed. Why? *Because of the underlying structure of the system.* This causes the key agents involved to behave the way they do, regardless of superficial attempts to force them to behave otherwise. Until this deeper layer of the problem is understood, the many diagnoses being offered by Al Gore and others will continue to be partial at best, and therefore unable to lead to an effective cure.

The Solution

The early chapters of *The Assault on Reason* go into considerable detail about how the assault is being achieved. An outstanding example is Gore's chapter on The Politics of Fear. As mentioned earlier on page 42, this could just as well have been named The Politics of Pushing the Fear Hot Button. The chapter on The Politics of Wealth is also an eye opener. But after that the middle chapters become little more than a long passionate listing of the sins of the George W. Bush administration. As Richard Ackerman explains in his review of the book: (Italics added)

> "This is unfortunate for two reasons. One is that this meaty diatribe in the midst of a book about reason is going to be the main section that many partisans seize upon. The second, and more important, as I stated in my opening paragraph, is that *it is factual yet irrelevant.* The entire section could have been replaced with a reference saying 'See factually documented issues with current executive in (selected list of books, articles, and websites)' with Gore's formidable intellectual energies then going into *analysis*.
>
> "*Reason is not about listing facts, it is about analyzing them to come to useful conclusions.* Yes, all these things happened. The war. The Kyoto situation. But WHY? Why aren't the American people reacting? Why aren't the American people informed? Why do they fear the wrong things and draw the wrong conclusions? Why are the people in power making certain choices? *Unless we can understand why people are doing things, we can't start to develop a strategy to address the situation.*" [64]

Without asking WHY questions like these and analyzing facts instead of merely listing them, problem solvers cannot arrive at a sound diagnosis. Without that, a correct solution is impossible.

Analysis is breaking a problem down into smaller problems so they can be solved individually, and using those conclusions to develop a comprehensive model explaining why the problem occurs and how the system would respond to proposed solutions. This has not been done here. Instead, we have a loosely constructed informal model of the system's behavior. The model is so loose and informal it's never listed or diagramed. *But the complexity of the problem Gore is addressing requires a formal model for correct analysis.* For problems whose dynamic behavior is difficult to fathom, this is best done with simulation models, as demonstrated in this book and the many models in Al Gore's *An Inconvenient Truth* that scientists have used to better understand and predict various aspects of climate change.

Without construction of a formal physical model, there is a tendency for problem solvers to build simplistic mental models and jump to simplistic conclusions. The chapter on The Politics of Fear is excellent. I learned a lot about why fear works so well as a manipulative strategy. But where does the use of fear fit into Gore's analysis model? It's hard to say, other than fear works, so it's used a lot by those who will do anything to achieve their ends. Where do the symptoms of less truth and more falsehood fit into Gore's model? Again, it's hard to say, other than there needs to be more truth and less falsehood if we are to solve the problem, and that control of television makes successful transmission of falsehoods easier and cheaper.

Contrast this to the crisp clarity of the Dueling Loops model. In it, fear is one of the five main types of false memes employed by corrupt politicians to win the race to the bottom. Why political falsehoods are preferred to the truth is explained by the model in detail. It's because the size of false memes can be inflated, but the size of the true memes cannot. *Where are powerful insights like these in the intuitive, informal models of The Assault on Reason and other analyses?* They are seldom found, due to the intrinsic weakness of the informal approach.

The informal analysis of the middle chapters leads to the solution presented in the next to the last chapter, on A Well Connected Citizenry. Gore carefully builds to the solution with this line of reasoning: (Italics added)

> "Many Americans now feel that our government is unresponsive and that no one in a position of power listens to or cares what they think. They feel *disconnected from democracy.* (p245)
>
> "I believe that the viability of democracy depends upon the openness, reliability, appropriateness, responsiveness, and two-way nature of the communications environment. After all, democracy depends upon the regular sending and receiving of signals—not only between

the people and those who aspire to be their elected representatives, but also among the people themselves. *It is the connection of each individual to the national government that is the key."* (p248)

Notice how this argument has drifted away from the earlier one that the use of expensive TV ads is the problem to solve. Now the feeling of being "disconnected from democracy" is the problem to solve. The two are related, but the diagnosis has morphed into something uncomfortably vague. Argument drift is common in informal models.

The new diagnosis leads, as you might expect, to a cure based on establishing a feeling of connection. It needs to be two-way, instead of the one-way of television. It needs to eliminate the "alienation of Americans from the democratic process." It needs to provide alienated citizens with "an effective way to communicate their ideas to others." Thus: (Bolding added. Italics are in the original.)

> "The remedy for what ails our democracy is not simply better education (as important as that is) or civic education (as important as that can be), but **the re-establishment of a genuine democratic discourse** in which individuals can participate in a meaningful way—a conversation of democracy in which meritorious ideas and opinions from individuals do, in fact, evoke a meaningful response.
>
> "And in today's world, that means recognizing that it's impossible to have a well-informed citizenry without having a well-*connected* citizenry. While education remains important, it is now connection that is the key. A well-connected citizenry is made up of men and women who discuss and debate ideas and issues among themselves and who constantly test the validity of the information and impressions they receive from one another—as well as the ones they receive from their government." (p254)

The Solution Examined

The core of this solution is "a genuine democratic discourse." I agree this is necessary, and I applaud Al Gore for explaining why in such detail. But isn't lack of such discourse merely just one more symptom of an unhealthy democracy, just as excessive deception, rampant cronyism, and unnecessary wars are also symptoms?

Thus this is a symptomatic solution. It attempts to treat the symptoms directly. It's like dipping a patient with a fever into a cold bath to lower the fever. This may work, but it will be temporary and will fail to arrest the underlying cause of the fever. Furthermore, if the patient consists of 300 million people, it's going to take a lot of effort to dip the patient.

Another way to describe this solution is it pushes on an intuitively attractive but low leverage point. Why do people use low leverage points again and again? The founder of the field of system dynamics, Jay Forrester, offers this explanation: (Italics and bolding added)

> "Social systems are inherently insensitive to most policy changes that people select in an effort to alter behavior. In fact, a *social system draws attention to the very points at which an attempt to intervene will fail.* Human experience, which has been developed from contact with simple systems, leads us to look close to the symptoms of trouble for a cause. But when we look, we are misled because the social system presents us with *an apparent cause that is plausible* according to the lessons we have learned from simple systems, although this apparent cause is usually a **coincident occurrence** that, like the trouble symptom itself, is being produced by the feedback loop dynamics of a larger system." [65]

Thus lack of a genuine democratic discourse is a coincident occurrence, rather than the root cause of the problem. Solutions focusing on coincident occurrences cannot solve problems, because they are not treating root causes.

Let's examine the lack of a genuine democratic discourse from the perspective of the Dueling Loops. The lack of democratic discourse would be one more sign that the truth, modeled as true memes, is not flowing as strongly as it should be. In other words, the race to the top is too weak to deliver on the promise of democracy.

But suppose you didn't know that the Dueling Loops model existed. You would naturally assume that pushing for more true discourse would solve the problem. But as the model shows, this would be pushing on the low leverage point of "more of the truth." *Pushing there will not work, due to the inherent advantage of the race to the bottom, and the fact that progressives do not have the force, in terms of numbers, money, and influence, to make pushing on low leverage points work.*

This would be no surprise to Jay Forrester, who knew "a social system draws attention to the very points at which an attempt to intervene will fail." And it should be no surprise to anyone who has studied the Dueling Loops.

The Second Edition of the Assault on Reason

The Assault on Reason contains tremendous insights on where democracy is heading and some of the reasons why. But it falls short on its self-imposed mission to "to figure out what has gone wrong in our democracy," which is the diagnosis, "and how we can fix it," which is the solution.

What might happen if Al Gore, working with an experienced modeler, used the Dueling Loops model as a starting point as he was writing the second edition of his book? His central premise that reason is under assault would not change, because this is the same as saying that the race to the top, where reason and truth prevail over deception, is currently not the dominant loop.

The superlative chapter on The Politics of Fear would change little. But it would probably be recast as one of the five main types of deception, instead of a popular strategy.

And so on. As the book's story unfolded, so would the formal model Al was building to analytically state his case. With each chapter his model would grow a little more, allowing him to stand on solid ground as each new premise and conclusion about the real world entered the artificial world of his simulation model. The narrative and the model would slowly build to a double climax: first the diagnosis, and then the solution. The main characters would be the social agents, strategies, and memes whose behavior he was modeling. The twists in the plot would revolve around construction of the model and the series of scenarios explored by running it. The book's thrilling moments would come when one force vanquished another, as loop dominance shifted as predicted or not, as insights built into the model's structure were tested and proven to be true or false.

While it probably would not include as much detail as the book you are reading right now, the second edition would provide a sufficient series of diagrams to paint the overall model, and the detail necessary to logically prove the particular line of analysis that Gore chose to follow.

Most importantly, compared to the first iteration of *The Assault on Reason*, the second would come to entirely different conclusions. And since Al Gore knows quite a bit more about political systems than the humble author of this book, he would probably take his model, diagnosis, and solution considerably further than this short book has been able to go.

Chapter 12

Taking Up Where Limits to Growth Left Off

W HAT WOULD BE THE BEST WAY TO APPLY THESE CON-
CEPTS TO A TANGIBLE WORK EFFORT, one that would make the
difference? If we are serious and believe in what this book has presented so
far, then we could hit the ground running if we launched a project taking up
where Limits to Growth left off.

From the viewpoint of the process we are executing, the Dueling Loops
are the *tentative diagnosis*. They explain why the human system is exhibiting
such strong change resistance. If this diagnosis is correct, then the sustainability problem is mostly solved, because in difficult cases of system dysfunction
the diagnosis is usually the hardest part. Once the reason a system is misbe-
having is known, the system can be properly treated. This is as true for social
systems as it is for biological or mechanical systems.

The System Improvement Process consists of these four main steps:

1. **Problem definition** (also known as problem identification)

2. **System understanding** (diagnosis)

3. **Solution convergence** (treatment plan development)

4. **Implementation** (treatment plan implementation)

Using the metaphor of modern medicine, the second step is the diagnosis,
the third step is development of the treatment plan, and the fourth step is implementing the treatment plan. Like the Scientific Method, it is a simple process. And it is an effective process, but only if all steps are performed well.

Society has yet to perform the diagnosis step. Instead, due to lack of a
process that fits the problem, it has skipped diagnosis entirely and rushed on
to the third and fourth steps with a vast collection of intuitively derived solutions. Some have worked, but only on easy problems such as local pollution.
The more difficult problems of climate change, deforestation, topsoil loss,
innumerable types of pollution, and many more remain flagrantly unsolved.

Things will remain this way until society performs the diagnosis step.
Let's examine how this could be done.

The Diagnostic Project: A Conceptual Proposal

In 1972 the Limits to Growth (LTG) project and book performed the first step of problem definition by correctly identifying the global environmental sustainability problem. The project identified the problem so well the book became the best selling environmental book of all time. But the book only performed the first step.

It appears possible to take up where LTG left off and duplicate its phenomenal success by duplicating what made LTG so successful. The key success **inputs** appear to have been: (1) The use of the right tool, system dynamics, to make the core analysis and argument, (2) A conceptual breakthrough by "seeing" certain system structures and emergent behaviors that had never been identified before, (3) Starting from a preliminary first pass at the project with the World2 model created by Professor Jay Forrester of MIT, (4) A highly qualified, well managed team, (5) A project sponsor in the form of The Club of Rome, and (6) Adequate project funding with a grant from the Volkswagen Foundation.

The practice of medical diagnosis did not emerge until the early 1900s, when William Osler, a Canadian physician, first enunciated the principles of the diagnosis and treatment of disease. As Osler saw it, the functions of a physician were to identify disease and its manifestations, understand its mechanisms, and determine how it may be prevented and cured. For his students he believed the best textbook was the patient himself.

The Oslerian ideal continues today. The basis of a doctor's strategy is "What disease does this patient have and what is the best way to treat it?" Osler is remembered for saying "If you listen carefully to the patient they will tell you the diagnosis." [66]

The System Improvement Process considers listening carefully to the patient to be the same as the System Understanding step. The only difference is the sustainability problem is like an infant who cannot yet talk—the patient cannot tell us anything directly. It is up to system analysts to extract that information indirectly, through modeling, measurement, and experimentation.

We would like to make a very simple argument: the phenomenal success of the Limits to Growth project is reproducible. It can be replicated if we can understand exactly why it succeeded and that pattern can be applied to the next step.

The purpose of the diagnostic project is to perform the diagnostic step. This can be done if all six of the above inputs are present plus a new one: a formal problem solving process that fits the problem. If the diagnostic step is performed correctly, treating the patient will be relatively straightforward, because the patient will respond in a predictable manner.

A well known principle of problem solving is that if two problems are similar, in terms of the inputs required for success, then solving a problem becomes a matter of providing the right inputs. From a project manager's point of view, problem identification and diagnosis are such similar projects that they have the same inputs. The lone exception is that since diagnosis is approximately an order of magnitude more difficult than problem identification, it requires a formal process.

LTG solved the problem of what is the overall problem? How real and serious is it? What happens if we don't solve it? While these questions may look trivial today, they were difficult and totally unanswered before LTG.

The diagnostic project will find the root causes of why the human system is unable to self-correct in time to avoid catastrophe. In other words, why is there such strong systemic change resistance?

The Phenomenon of Change Resistance

Notice how we have framed the problem. Calling it a change resistance problem runs against the conventional problem definition, which is more like "What do we need to do to be sustainable? What are the proper practices everyone should follow?" The weakness in this viewpoint is that we already know what to do to be sustainable. The technologies, practices, and lines of further research are already well known. The real problem is most of the world just doesn't want to follow the proper practices and technologies required to live sustainably.

This point deserves some emphasis, because it holds the key to taking up where Limits to Growth left off.

As explained earlier, the social side of the problem is the crux. Society knows HOW to be sustainable. It just doesn't want to DO it. This is known as change resistance. Because it exists deep within the human system and is global, it is systemic. Until the systemic change resistance part of the sustainability problem is solved, it does little good to plead, over and over again, that we must be sustainable. Better is to find the root cause of change resistance and direct your efforts there.

Below are several examples of change resistance. Some are direct and some are indirect. The latter is generally preferred, because it has more leverage and is less likely to cause a backlash.

After Silent Spring was first published in 1962, "Not surprisingly, both the book and its author… met with considerable resistance from those who were profiting from pollution. Major chemical companies tried to suppress *Silent Spring*, and when excerpts appeared in *The New Yorker*, a chorus of voices immediately accused Carson of being hysterical and extremist—charges still heard today whenever anyone questions those whose financial well-being depends on maintaining the environmental status quo." [67]

"Corporations have long utilized think tanks and a few dissident scientists to cast doubts on the existence and magnitude of various environmental problems, including global warming, ozone depletion, and species extinction. This strategy is aimed at crippling the impetus for government action to solve these problems, action which might adversely affect corporate profits. … The think tanks have been so successful at clouding the scientific picture of greenhouse warming and providing an excuse for corporations and the politicians they support *that they have managed to thwart the implementation of effective greenhouse gas reduction strategies by governments in the English speaking world.*" [68] (Italics added)

"The defenders of business-as-usual on climate change began twenty years ago by telling us that concern about global warming was not scientifically justified. A decade later they said yes, concern is justified, but we have ample time to solve the problem. Now they are saying it is too late to prevent major climate change, and our best strategy is to adapt to it." [69]

Here is a more in depth example: (Italics added)

"As late as 1989, the tax on wine remained a constant one cent a gallon. But California was changing. The California Highway Patrol pushed the state to see alcohol as a public safety issue. Mothers Against Drunk Driving (MADD) turned personal grief into political mobilization.

"In the late 1980s, a broad coalition of groups organized to pressure the [California] state legislature to impose a nickel-a-drink tax on bars and restaurants, with revenues earmarked for trauma centers, law enforcement, alcoholism prevention and treatment." Polls showed "that 73% of Californians supported such a tax."

"Responding immediately, liquor industry leaders held emergency meetings to plot a counter strategy. The president of the California Wine Association called Proposition 134 'the most serious threat to this country since Prohibition,' and an industry newsletter reported that the industry would spend 'whatever is necessary' to defeat the tax. Led by donations from Seagram & Sons and Guinness Corporation, the industry committed *an unprecedented $38 million to oppose the nickel-a-drink tax initiative.* Attack ads were drafted, a one-penny-a-drink counter initiative was launched in order to muddy the waters, and an industry front group, Taxpayers for Common Sense, was created in the offices of the liquor executive.

"The combination of negative advertising, counter initiatives, front groups, and an overwhelming financial advantage proved effective. In the crucial area of broadcast advertising, the balance of resources was not even close. *While the liquor industry spent $18 million on ads that slammed the nickel-a-drink initiative, proponents had only $40,000 with which to counter them.* On election day, confused voters rejected both the citizen initiative and the industry alternative." [70]

Then there is this outrageous example, where a member of the New Dominant Life Form brought the state of Montana to its knees in 1903, with "a brutal tactic known as The Great Shutdown:"

"Consider the state of Montana, which for nearly a century was run as a virtual colony by the aptly named Anaconda Copper Company. The company had a tradition of corruption and hardball tactics. At one point, displeased with the decision of a state judge in favor of one of its rivals, Anaconda shut down all its mines and smelters in the state for three weeks, cutting off thousands of workers from their paychecks, until the governor called a special session of the legislature to pass a new bill that resolved the dispute in favor of the company." [71]

These are stunning examples. They show how unstoppable a powerful opponent can be if they want to preserve the status quo so that it remains in their favor. Such opponents can stall or block solutions to any problem, as long as they see that as in their best interests.

Returning to the sustainability problem, think of civilization as a sick patient who doesn't want to take his medicine. He has a million and one fallacious reasons why not: "It tastes bad. It's too expensive. I'm not really that sick. Those tests are not conclusive." And so on. But what happens if he doesn't take the medicine? The patient dies.

The real question is WHY is the patient so strongly resisting changing his behavior to one that's good for him? In other words, what is the root cause of his resistance?

On page 25 we quoted the third edition of *Limits to Growth* as saying:

"...humanity has largely squandered the past 30 years..."

That humanity has squandered the past 30 years is proof positive that systemic change resistance is present. Why it is present and how to overcome it is *the* problem to solve, because once it is overcome, the human system will now inherently want to be sustainable. After that the problem becomes a simple matter of developing, selecting, and implementing the most efficient practices to live sustainably. *Once change resistance is overcome the rest of the problem is several orders of magnitude easier to solve, because the system is now self-managing and thus desperately wants to solve the problem.*

This is all perfectly normal for difficult social problems. In fact, change resistance is almost always present in difficult social problems, because if it wasn't present they would be easy. Therefore any serious approach to solving the problem must consider change resistance.

I am not alone in this observation. Professor Jay Forrester also noticed the phenomenon of change resistance, though from a slightly different perspective. He created the World1 and World2 models which became the World3 model in LTG. In 1971 Forrester published World2 in *World Dynamics*, a small book with 137 pages. In 1972 the first edition of *The Limits to Growth* was published. A year later, in 1973, Forrester released a second edition of *World Dynamics*. It contained a new five page chapter at the end called *Postscript—Physical Versus Social Limits*.

In general the new chapter pointed out that problem solvers were too enamored with the technical side of the problem, and were thus ignoring the social side. Forrester described the error this way:

> "Debate about future implications of economic growth has focused almost entirely on physical limits and the role of technology in pushing back physical limits. But to concentrate on physical limits is to ignore the increasingly important social limits to growth.
>
> "[The first edition of] *World Dynamics* has contributed to a misplaced emphasis on physical limits by understating the importance of the crowding mode of the model in Section 4.4. *The Limits to Growth* also veered away from social and political factors to stress the more tangible physical aspects of the world environment. This chapter has been added to the original text of *World Dynamics* to set the relationship between physical limits, growth, and social limits in better perspective.
>
> "The debate about growth has centered on resources, pollution, and agriculture. But the most important issue is not the ability of technology to continue pushing back the physical limits. The question can be better stated, 'Assuming technology can continue to push back the physical limits of the earth, [why] should society want to do so?'
>
> "Relying on technology to solve the problems created by growth is to evade the question of how to slow growth." [72]

The new chapter included a small model (only 8 nodes) roughing out how the physical and social limits of the system were related. The chapter viewed the social side of the problem as one of the need to consider "self restraint," "social stress," and "social limits," rather than our concept of the need to overcome change resistance. But Forrester did implore the reader to not ignore the social side of the problem, because if it is not addressed then the problem remains insolvable.

We feel analyzing, understanding, and overcoming change resistance is the crux of solving the sustainability problem. But this proposition, viewpoint, paradigm, or whatever you wish to call it seems nearly impossible for most problem solvers to grasp. Even if they do they give it little emphasis, which means they have not truly grasped its critical importance.

Even the third edition of *Limits to Growth* slipped into this pattern, as shown in this passage from the chapter on Tools for the Transition to Sustainability: (Italics added)

"...*systems strongly resist change* in their information flows, especially in their rules and goals. It is not surprising that those who benefit from the current system actively oppose such revision. Entrenched political, economic, and religious cliques can constrain almost entirely the attempts of an individual or small group to operate by different rules or to attain goals different from those sanctioned by the system. *Innovators* can be ignored, marginalized, ridiculed, and denied promotions or resources or public voices. They can be literally or figuratively snuffed out."

There it is, "systems strongly resist change." But where did this thread of thought go? It slid the wrong way, into how the system would slap back at and resist the innovators, as you can see in the ending of the above paragraph. The next paragraph slid even further:

"Only *innovators*, however—by perceiving the need for new information, rules, and goals, communicating about them, and trying them out—can make the changes that transform systems. This important point is expressed clearly in a quote that is widely attributed to Margaret Mead: *"Never deny the power of a small group of committed individuals to change the world. Indeed it is the only thing that ever has."*

The logical thread of "systems strongly resist change" has disappeared, replaced by an inspirational call to innovators to keep trying. While this is commendable and necessary, it is hardly crucial, because for the last 35 years we have had millions of highly committed environmentalists working on the problem, many for their entire careers. The need to keep trying is not the bottleneck. Something else must be.

Let's imagine what that something else might be by rewriting the second paragraph:

"Only *innovators*, however—by perceiving the need to crack the wall of systemic change resistance preventing changing the rules and goals of the system—will find a way to do exactly that. This important point, *that overcoming change resistance is the innovative crux*, will drive innovators wild, and bedevil them with its near intractable difficulty, until they find a way to solve that part of the problem."

The call to inspiration in the original second paragraph pushes on a high leverage point for easy problems, but a low leverage point for difficult problems. High change resistance is what makes a social problem difficult. *It fol-*

lows that for innovators to crack the wall of change resistance, their innovation must come in the form of finding the true high leverage points in the system. Only then will the rest of the millions of problem solvers have points to push on that will work.

This leaves us with a decisive question: How will we find those high leverage points?

The Key to Success Is a Process That Fits the Problem

That question carries us to the key to a successful diagnosis, which will find the high leverage points. *Our work must be guided by a process that fits the problem.* Doctors have one. They are trained on it in med school and residency. Scientists have one. It is the Scientific Method. Without it they would be as lost as a ship at sea without a navigation system. And finally, business managers have a process that fits the problem of how to maximize profits. It's based on double entry accounting and the principles of performance feedback and sound financial management.

But where is a process that fits the problem (and addresses elements like change resistance, low leverage points, and high leverage points) to be found in environmentalism? There is none. Every activist and every organization takes a different informal, intuitive approach. One result is the old joke "If you get ten environmentalists in a room, how many opinions will you get on how to solve a problem? Eleven." The other result is not so humorous. It is endless solution failure on difficult problems.

Up until now, environmentalists have been able to get by without a formal process because, like most problem solvers, they tackled the easier problems first. Once the sustainability problem was spotted, it was much easier to solve local problems than global ones. It was also much easier to solve problems whose consequences appear sooner rather than later. This explains, for example, why water source pollution problems have been so much easier to solve than the climate change problem. If environmentalists want to become capable of solving difficult problems, they will need a process capable of doing that. Neither science nor the business world knows any other way.

Environmentalists have made notable progress. But by not using a written, comprehensive, formal process, they have nothing explicit to improve. *The result is they have been unable to continuously improve a common process until it is good enough to solve the complete problem.* [73] Another consequence is they have been oblivious to the real problem to solve first: change resistance.

A process that fits the sustainability problem has been developed. This is the System Improvement Process (SIP). It is a simple, generic, analytical process designed to apply to all complex social system problems. It has four main steps. The first main step defines the overall problem. The process then decomposes the overall problem into three subproblems, and uses main steps 2, 3, and 4 to solve each of them. The three subproblems are:

1. **How to overcome change resistance** to adopting proposed solutions, also called proper practices.

2. **How to achieve proper coupling**. (Defined on page 23) This moves the system from its present state to the goal state. This occurs due to adoption of the proper practices, as a result of overcoming change resistance and resolving the root causes of improper coupling.

3. **How to avoid excessive model drift** – The overall solution must keep the system in the goal state. If the solution drifts too far from what's needed the problem will recur.

The subproblems are sequential. Change resistance must be overcome so that proper coupling can be implemented. Model drift must be eliminated to prevent overall problem recurrence.

The *goal state* of the system occurs when problem symptoms are reduced to acceptable levels. If the system is staying in the goal state or is moving there in time, the problem is considered solved. *In the sustainability problem, moving to the goal state is the same as the proper coupling of the human system to the greater system it lives within, the biosphere, so that the health of the two systems is automatically maintained indefinitely.* High quality proper coupling requires the right feedback loops to exist, so that the solution is self-managing and reasonably optimal. The right feedback loops will cause the right decisions, technologies, and proper practices to appear and be used.

The popular conception of the word "solution" means proper coupling. However, as this book argues *the real problem is how to overcome systemic change resistance*. That is the crux of the problem.

In problem solving jargon, a **solution space** is all possible solutions. The System Improvement Process provides an extremely efficient means of searching a large and unknown solution space for a solution that will work. The reduction of millions of possible solutions to ones that will actually work is known as Solution Convergence, which must be preceded by System Understanding so that convergence happens quickly and correctly.

Here's an outline of the **System Improvement Process**:

1. Problem Definition – What is the problem? This is defined in terms of the goal state versus the present state of the system.

2. System Understanding – Why are the three subproblems occurring?

2.1 Why is there such strong resistance to adopting the solution?

2.2 Why is the system not naturally in the goal state?

2.3 Why is the system not staying in the goal state?

3. Solution Convergence – How can the three subproblems be solved?

3.1 How can adoption resistance to the solution be overcome?

3.2 How can we move the system to the goal state?

3.3 How can we keep the system in the goal state?

4. Implementation – Once a solution is found, the three subproblems are solved in this order:

4.1 Overcome resistance to solution adoption.

4.2 Move from the present state to the goal state.

4.3 Stay in the goal state indefinitely.

Limits to Growth performed step 1, Problem Definition. The diagnostic project will perform step 2.1, Why is there such strong resistance to adopting the solution? Full diagnosis would also require steps 2.2 and 2.3. However, we hypothesize that change resistance is the bottleneck. Once the subproblem of change resistance is diagnosed and then overcome in steps 3.1 and 4.1, the human system will then automatically and aggressively seek to solve the rest of the problem. This will be a rather pleasant change.

Note the clarity and focus these steps give all work effort. In a formal process like SIP, each process step is a distinct, well defined problem to solve. Formal processes make work more efficient by redefining one big job into lots of much smaller and hence easier to perform little jobs. *When applied to difficult complex system problems, this decomposition is so powerful it can routinely transform a problem from insolvable to solvable.* This is probably the case here. The chronic absence of a process like SIP is, in our humble opinion, the main reason "humanity has largely squandered the past 30 years."

The System Improvement Process thus becomes the foundation for a successful diagnosis. It is the most important success input by far, because it is the driver for everything else.

You may have noticed our resolute adherence to a formal process that fits the problem like a glove. The two principles we are following are:

1. The process must fit the problem.

2. The more difficult the problem, the better the process must be.

Following these rules leads to a powerful emergent property: *problem solving becomes a matter of methodical, relentless process execution.* Or, as Toyota has concluded from decades of success, "The right process will produce the right results." To those who have never approached problem solving this way, this is a completely new paradigm.

The remainder of this proposal will take a brief look at the other key success inputs, as listed at the beginning of this chapter. Following this discussion, we will draw the conclusion that the phenomenal success of LTG was not a fluke. We can make it happen again.

Key Success Input 1: The use of the right tool, system dynamics, to make the core analysis and argument

As explained in chapter one, **system dynamics** is a modeling tool. Its purpose is to more deeply and correctly understanding the dynamic behavior of social systems.

System dynamics is *the* essential tool for identifying, analyzing, and solving difficult social system problems at the tactical level. The tool rose to instant international prominence with publication of *The Limits to Growth* in 1972. Right there on dozens of pages were the model diagrams and system behavior graphs that so persuasively showed the cliff civilization was marching toward, and what would probably happen if certain scenarios played out. These scenarios awakened the world to an unexpected new truth.

Some readers agreed. Others did not, and tried to shoot the messenger, with a blaze of accusations and critiques. But none were able to shoot the tool that, in the hands of experts, provided the analysis and argument. The end result is that today LTG's reasoning and key conclusions remain as irrefutable and useful as ever, though there is still an ample supply of naysayers and nitpickers.

The success of the LTG project is as inconceivable without the use of system dynamics as science is without the Scientific Method. It was the right new tool at the right time for the right new problem. There is little reason to doubt that, if this tool is focused correctly on the diagnostic step, it can do it again.

Key Success Input 2: A conceptual breakthrough by "seeing" certain system structures and emergent behaviors that had never been identified before

Forrester used the tool of system dynamics to "see" system structures that no one had ever seen before. Once the structures were defined, running the model showed how the system would behave, within a broad range due to the way all models are a simplified representation of reality. And then, once Forrester had created the first two iterations of the model, the LTG team was able to refine it still further into World3, giving the model even greater explanatory and predictive power.

The conceptual breakthrough was that people could now talk and think at a whole new level of understanding. This elevated debate from an intuitive, haphazard level to an analytical level, where terms like "limits to growth," "exponential growth," and "unsustainable" now had clear meanings and consequences. Environmentalism had grown up overnight. It now had the language and theory that every new field needs, both to be taken seriously and to make serious contributions.

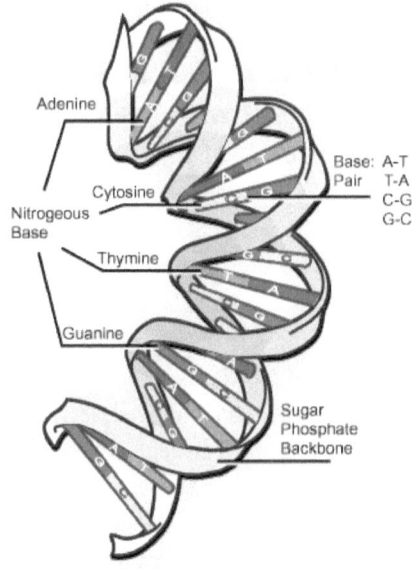

The conceptual breakthrough of this proposal centers on finally "seeing" the structure of the crux of the problem. If we can do this then the problem is 80% solved. Like the LTG project or many other scientific problems, we need to "see" certain system structures that no one has ever seen before. Once we can see those structures, the world will be stunned by how much more they can explain and how the sustainability problem is now an order of magnitude easier to solve. [74]

Once Crick, Watson, Wilkins, and Franklin discovered the structure of DNA in 1953, by research and building molecular models until they found one that worked, biology exploded into a frenzy of progress. Forensic identification via DNA, anthropological uses of DNA, medical applications based on knowledge of DNA, and genetic engineering have transformed our world.

And it was all because scientists could now "see" the very essence of what it was they needed to work with: the structure of the code of life.

From this follows the critical need for a formal process emphasizing the change resistance aspect of the problem. From this also follows the need to dig deep into the problem to find and thoroughly understand the root cause of change resistance, which requires the tools of system dynamics and memetics.

The concept that overcoming systemic change resistance is the crux of the problem is a subtle idea and a new paradigm. Very few environmentalists, scientists, and politicians, even those at the national and international level, see their world this way. They are instead committed to the paradigm of Classic Activism, which sees the problem very differently. (See the Glossary at Thwink.org for what Classic Activism is.)

It will not be easy to change the mental models of classic activists, but from the perspective of the System Improvement Process, that is merely another part of the problem to solve.

Key Success Input 3: Starting from a preliminary first pass at the project with the World2 model created by Professor Jay Forrester of MIT

The LTG project would have never occurred without Forrester's World2 model. It appears that no one else in the world had his modeling skill. Of course he was the inventor of the tool, so one would expect him to be proficient. But the tool was so new, and by today's standards so immature, that no one else came close to his combination of high proficiency and interest in solving social problems.

Today that has changed, but only partially. System dynamics is now taught as a standard college course. Superlative textbooks like John Sterman's *Business Dynamics: Systems Thinking and Modeling for a Complex World* exist, allowing serious students to teach themselves (as I did) or take a college course. Easy to use, highly mature software programs like Vensim allow almost anyone with reasonable computer skills to pick up another skill: how to model dynamic problems. However, as easy as the tool is to learn, applying it well is just as hard as ever. As a result, there are very few adequate first passes at the change resistance side of the sustainability problem that could serve the same breakthrough role that Forrester's World2 model played.

One that could possibly fit this need is *The Dueling Loops of the Political Powerplace* model. This model presents the novel hypotheses that the Dueling Loops are not only the root cause structure behind systemic change resistance to solving the sustainability problem. They are also the root cause behind widespread, prolonged resistance to solving nearly all problems whose solution would benefit the common good of all, rather than the few.

The Dueling Loops model could be called the *tentative diagnosis*. The model has reached the point where experimentation is necessary to confirm the diagnosis and calibrate the model so that leverage point behavior can be predicted with a high level of confidence. *Thus a fast track to a diagnosis exists: confirm the tentative one.*

Of additional interest is the Dueling Loops model was developed using the System Improvement Process. Thus the model and documented simulation runs emphasize the attractive but low leverage points that problem solvers are presently pushing on, and the high leverage points they need to push on instead to solve the change resistance problem. This allows a diagnosis of such depth that the next step, treating the patient, should be relatively easy, because the model tells us how the system is likely to respond to the most promising solutions. The model can also be used as the first iteration of the solution development (treatment) model, which is one of its intended roles.

Other models probably exist that could help play the preliminary first pass role. At the very least the Dueling Loops model, due to its extreme novelty, may jolt the modelers on this project into the innovations needed for a breakthrough.

Little known is the fact that in the right hands, good first pass models can be created quickly. For example, here's how long it took Professor Jay Forrester to create the World2 model: [75]

"For those who may be unfamiliar with events following the publication of *World Dynamics*, a little history may be of interest. The model in the book was the product of only two Saturdays of work. The book [itself] took another four months to write, edit, and get all the computer runs onto consistent scales."

Key Success Input 4:
A highly qualified, well managed team

More than anything else, the LTG project centered on the new tool of systems dynamics. The inventor of system dynamics, Jay Forrester, and the world's preeminent educational institution on teaching system dynamics, MIT, were involved from the start. This made it relatively easy to hand pick the most talented team in the world. The modelers were all protégés of Forrester. The project manager, co-modeler, and co-author was Dennis Meadows, who has a PHD in management from MIT. The team was so MIT centered it was frequently called "the MIT team."

Now then, how are we going to put together an equally highly qualified, well managed team?

This question was discussed with Phillip Bangerter, Steve Gale, and Andrew Murphy of Hatch (www.hatch.ca) on December 14, 2006. Hatch is a global engineering consultancy. Phillip's opinion was that this project, aside from its slightly academic aspects, is a typical challenging engineering project. Hatch and many other consultancies could manage a project like this using the same highly refined processes they apply to other similar projects. They could also provide the project team a "home" (at least in the startup phase), with all the infrastructure and connections that would make the project much more efficient. This is identical to the home that MIT gave the LTG team. This could be the leading option for team management.

As for team staffing, once we have a proposal that demonstrates this project will take up where the last team left off and will be just as likely to hit another home run, then it should be possible to interest MIT and other leading institutions into helping to assemble another world class team. The key will be to show we know exactly what caused the success of the LTG project, and we know how to use that knowledge to replicate its success. This will take some effort to produce. The results will go far beyond this first version of the proposal.

Let's explore how we can attract a world class team.

Once you understand it, the key advantage to this proposal is *thinking at the meta level*. It prioritizes the problem solving process as the first thing to get right. Conventional approaches begin with the implicit assumption that the same methods we have always used will suffice. Nowhere in traditional methods will you find a process that fits the problem.

This results in very low process efficiency. We can say this with some assurance, using the chapter on *An Assessment of Process Maturity* in the manuscript to *Analytical Activism* at Thwink.org. The table summarizing the findings of the assessment is on the next page. This chapter rates ten representative environmental organizations on process maturity, including the largest in the world. Process maturity is so low that eight out of ten scored below 500 on a scale of zero to 10,000. Only two scored moderately well, at about 5,000: The Nature Conservancy and the European Union Environmental Directorate General. The key finding was that none are thinking at the meta level, with one exception: The Nature Conservancy. However their process, Design for Conservation, only fits a small portion of the sustainability problem. Still, the Conservancy is a fine example of what is possible. But even the Conservancy's process scores only 4,489. A diagnostic team driven by the System Improvement Process would score about an 8,000. The chapter argues this would nearly guarantee mission success.

An Assessment of Process Maturity
Showing the dominance of Classic Activism and why that causes low mission success

The table is designed to assess process maturity for solving difficult environmental problems. The assessment was performed in 2006.

Only the weighted scores are shown. To calculate the raw scores, divide the weighted score by the element weight.

Raw scores for each key process element are assigned in this manner:

0 – Does not exist or not done
1 – Very low productivity
2- Slightly productive
3 – Moderately productive
4 – Highly productive
5 – World class

An underline means not applicable, with an automatic raw score of 3.

| Organizations | Key Process Elements | | | | | | | | | | | | | | Total score on a scale of 0 to 100 | Process Maturity Rating = Total score squared, on a scale of 0 to 10,000. | Overall mission success – Low, Medium, High |
| | Classic Activism | | | | | Analytical Activism | | | | | Problem Domain | | | | | | |
	1. Identify the problem	2. Find the proper practices (PP)	3. Tell the people the truth about the problem and the PP	4. If that fails, exhort and inspire people to support the PP	Weighted subtotal	5. Formal definition, mgt, and continuous improvement	6. A true analysis of the problem is performed	7. The Scientific Method is used to prove all key assumptions	8. Learning from past failures and successes is maximized	Weighted subtotal	9. The analysis centers on a social system structural analysis	10. Low and high leverage points have been found an tested	11. Why change resistance is so strong has been determined	Weighted subtotal			
Element weight	1	1	1	0	15	4	3	3	2	60	2	2	1	25	100	10,000	
Max weighted score	5	5	5	0	15	20	15	15	10	60	10	5	10	25	100	10,000	
1. Alliance for Climate Pro	3	3	5	0	11	0	0	0	4	4	0	0	0	0	15	225	L
2. Club of Rome	2	2	5	0	9	0	3	0	4	7	0	0	0	0	16	256	L
3. European Union Env DG	5	5	5	0	15	12	9	9	10	40	6	8	3	17	72	5,184	H
4. Natural Step	3	4	5	0	12	4	0	0	1	5	0	0	0	0	17	289	L
5. Natural Resources Def C	3	3	5	0	11	0	0	0	6	6	0	0	0	0	17	289	L
6. Nature Conservancy	5	5	5	0	15	20	12	12	8	52	0	0	0	0	67	4,489	M
7. Sierra Club	2	3	5	0	10	4	0	0	4	8	0	0	0	0	18	424	L
8. United Nations Env Prog	1	3	5	0	9	0	0	0	4	4	0	0	0	0	13	169	L
9. Union of Concerned Sc	2	5	5	0	12	0	0	0	2	2	0	0	0	0	14	196	L
10. World Resources Inst	2	5	5	0	11	0	0	0	6	6	0	0	0	0	18	424	L
(Solution Factory)	5	3	3	0	11	16	15	15	8	54	10	10	5	25	90	8,100	?

Once an organization becomes formal process driven, it has reached a new level of self-awareness. Because it is now aware of how it is thinking at the macro, organization wide level (the thinking is the process), it can continuously improve its own thinking. Over time this will cause process driven organizations to become better (smarter) than non process driven ones by several orders of magnitude, in terms of the difficulty of the problems they can solve, and the efficiency (speed and cost) in which they solve them.

The same applies to project teams. Furthermore, some engineers are becoming aware of the value of being process driven. The rise of Six Sigma in college curriculums and industry is but one example.

In my opinion it is the very best engineers, including modelers, who are attracted to high quality process driven companies and projects, because they want to work with the best technologies possible. Thus if we can present the project as one using a breakthrough process that is likely to produce breakthrough results, we should be able to attract a fine team.

Key Success Input 5:
A project sponsor in the form of the Club of Rome

A project sponsor is an entity who wants the project to succeed, and will move heaven and earth to make that happen. They do not necessarily directly manage, fund, house, or staff the project. After the quality of project management, the quality of the project sponsor is the most important key success input for project success.

The LTG project was fortunate to have the best possible sponsor: the Club of Rome. At the time the project began, there were very few organizations committed to making serious progress on the sustainability problem, because it was invisible except to a precocious few. Like attracts like, which caused a most fortuitous event: the formation and growth of the world's first global think tank centered on the world's biggest problems. It did not take them long to conclude that there was one problem that dwarfed all others: the global environmental sustainability problem.

Acting on that epic insight, the Club of Rome initiated a project to solve what they called The Predicament of Mankind. Once again like attracted like, causing the project to attract Jay Forrester and with him the rest of MIT. At this point the project had key success inputs 1, 2, 3, 4, and 5. Because the project had a good sponsor, number 6, adequate funding, came easily. All Eduard Pestel, a member of the Club's executive committee, had to do was ask his own foundation, the Volkswagen Foundation of Germany, to fund the entire project.

But times have changed. The Club of Rome is no longer a giant of change. In fact no environmental NGOs are. They are instead dwarfed by transnational corporations, who are the real giants. As explained earlier in this book, the modern corporation is now the most powerful life form on the planet. It is therefore potentially the most powerful sponsor—if we can find big successful ones who are already pro-sustainability visionaries.

Fortunately a few of these exist. It is several of them who will probably become the ideal sponsors of this project, possibly combined with a few of the very best environmental organizations so as to make certain aspects of the project, especially those coming after diagnosis, go faster and more efficiently.

Key Success Input 6: Adequate project funding with a grant from the Volkswagen Foundation

It took only a single source to adequately fund the LTG project. The money was easily obtained, because the project was so obviously grant worthy.

The same will happen to this project, if we can show that it is deserving of funding. This should be relatively easy because of the radical difference between this project and most others. This project is process driven. The others are not. It also uses root cause analysis and system dynamics to diagnose the root causes of systemic change resistance. The others lack this conception.

* * *

This completes the review of the key inputs of success.

The Limits to Growth book contained such breakthrough content that it sold itself. Yes, it was well written and marketed. But it was the book's extraordinary content that caused it to ultimately become the best selling environmental book of all time. Today, over 30 years and two more editions later, it has sold over 30 million copies. The next closest is *Silent Spring* with 10 million copies. [76]

The necessary preconditions for project success are implied by the process steps. The necessary precondition for problem identification is an undiscovered problem that must eventually be addressed. The precondition for a root cause diagnosis is the existence of a root cause of the problem's symptoms. This can be assumed to be the case for all social system problems whose solution is less than obvious.

While this chapter speaks of a single project for simplicity, the best approach is many Diagnostic Projects using constructive competition. This is a widely known best practice, like the use of competing design teams. Thus we hope to see not one but many Diagnostic Projects. Only one has to hit a home run to fully and correctly diagnose the problem. But two doubles and a triple would do the same.

Managing This as a Mega Scientific Frontier Project

Engineers know that successful project outcomes can be replicated if the inputs and preconditions are the same, within an allowable range of variation. That appears to be the case here.

Those who have managed difficult projects will instantly see the strong pattern of similarity between the Limits to Growth and Diagnostic projects. As this short chapter has explained, both projects have the same inputs at the broad brushstroke level. Both projects are also a step in the System Improvement Process and thus have the necessary precondition for the project to fit the step. This is sufficient logical proof that the LTG project was not a fluke. We can make it happen again, simply by approaching it as a typical cutting edge engineering project that requires the usual tight managerial controls to ensure success.

This is not your average highly challenging engineering project, however. It is much more difficult, by approximately an order of magnitude. This aspect must be formally managed or the project will fail. Let's examine this.

Looking ahead, the diagnosis step will require a large collection of significant scientific breakthroughs. This puts the project in a rarified class of projects that could be called **scientific frontier projects**. Examples are the first nuclear bomb project (the Manhattan Project) and the first man on the moon project (the Apollo Program). Both projects required several major and many minor scientific breakthroughs to succeed. This made them not just cutting edge but bleeding edge projects, because if any required discovery failed to be made, the entire project was jeopardized.

As big as the Manhattan Project and the Apollo Program were, they are dwarfed by the size, complexity, and difficulty of the global environmental sustainability problem. No project mankind has ever faced comes close, including decoding the human genome, building the Panama canal, and the seven wonders of the ancient world. It is a **mega scientific frontier project** of a scale that boggles the mind.

But it should not boggle the minds of engineers who realize that if they apply their tools methodically and correctly, the problem will yield quickly to solution.

Moving Forward with Social System Engineering

The particular type of engineering to be used is social system engineering. Like any other branch of systems engineering, this is the ability to design, build, modify, and repair a particular type of system, such as a manufacturing process or a corporate information system. In social system engineering the system we are interested in is social control models. Let's define this term.

A **model** is a simplified representation of reality. Models serve as references for decision making. A model may be physical or mental.

Models fall into several classic types: *Descriptive models* are data, such as maps and history. *Behavior models* describe how and why something behaves, such as physics, biology, and system dynamics simulations. *Control models* are built to allow control of the world around us, such as the principles of architecture or the rules followed to tame a wild horse. A control model is the set of rules needed to control the outcome of something.

A crucial type of control model is the **social control model**. *A social control model defines how a social unit runs itself.* Once a social control model is perfected, it can be used over and over. Examples of modern social control models are the ones used by families, school systems, countries, congregations, and corporations. Each has an unwritten and/or written set of rules that describe how the social unit should work. For example a legislative body follows the rules of a constitution and, during deliberations, follows Robert's Rules of Order or some other set of debate rules. The oldest social control model is probably the family.

From the viewpoint of solving the sustainability problem, the most important social control model is the one that global civilization is using to run itself. This is the model that's broken, because it is currently unable to achieve its goal of running civilization well. Thus the model is in the Model Crisis step of the Kuhn Cycle, as explained on page 184.

Currently the ability to reliably engineer social control models is nonexistent. None were ever really engineered. Instead, they evolved over long stretches of time, with too many periods where too many people suffered. This needs to change so that we can proactively solve the sustainability problem. This can only be done if the social control models involved are capable of aggressively and correctly solving the problem. How to fix these models so they can do that is the first challenge of the emerging field of social system engineering.

If we cannot mature this new field quickly, then we will be unable to apply the principles of engineering to the problem, and will be forced to fall back on what we're doing today, which is not working.

Notice our strategy. We are not advocating electing politicians who want to solve the sustainability problem. Nor are we trying to fix certain governments or international agencies. And we are not trying to ram through certain legislation to regulate certain environmental problems. *These would be symptomatic solutions.* They might give some short term success, but in the long run they would fail. Instead we must look deep within the system and resolve the root cause of why our social institutions have so badly failed us. Our findings are this:

> *Going beyond the root causes found in the Dueling Loops, the deeper root cause is the lack of ability to engineer our basic model of democratic government so that it can reliably achieve its goals.*

For the Diagnostic Project to correctly determine exactly why systemic change resistance is so strong, it needs to first perfect the new tool of social system engineering. Then the project can confidently say, "This is the root cause of change resistance. Right here is where the governmental social control model is broken. Fix it and the system will then automatically self-manage a solution to the sustainability problem, and many more problems like it."

Only with that generic insight will we be able to fix the many social control models around the world that are behaving so unsustainably. These are the many local, state, and national governments making up the 190 or so countries of the world. If we can fix them, then they in turn will create or fix the global social control models required to solve the problem in a coordinated manner, such as a United Nations and a United Nations Environmental Programme that actually work.

Avoiding the Trap that Got Us Where We Are Today

If you have studied the steps of the System Improvement Process as listed on page 169, a few burning questions may have arisen by now: Why stop at diagnosis? Why not target this project to the full solution of the change resistance part of the problem? Why not include all three SIP subproblems and execute the entire process?

This needs to be done. But this conceptual proposal has deliberately avoided doing it, because it is too easy to fall into a trap.

This is the **Jumping to Conclusions Trap**. Most people and organizations working on the global environmental sustainability problem fell into this trap long ago, without ever knowing it. From the perspective of the System Improvement Process, the trap occurs when problem solvers go straight from Problem Definition to Solution Convergence. *They have skipped step two, System Understanding.* By skipping this crucial step, the one where they should spend about 80% of their time, they have jumped to intuitively attractive but totally wrong conclusions about how to solve the problem, as the past 35 years have so forcibly demonstrated. They have essentially put the cart before the horse, and are mystified when the cart keeps running off the road.

This proposal seeks to rectify this error by saying let's back up and do step two right.

Then we can go ahead.

* * *

The Four Main Steps of SIP
1. Problem Definition
2. System Understanding
3. Solution Convergence
4. Implementation

Like the pyramids of Egypt and the cathedrals of Europe, it will take many years to do step two right and then go ahead. But once we do, and finally establish a dominant race to the top in the political powerplace, a whole new level of being becomes possible. *Because the truth has no higher master, a permanent race to the top, and all that can lead to, is now within our grasp.*

Chapter 13

The Tantalizing Potential of a Permanent Race to the Top

IMAGINE A WORLD WHERE ALL ELECTED LEADERS ARE MOTI-VATED TO DO THE BEST THEY POSSIBLY CAN for the people as a whole. The universal goal of all governments and politicians would be to optimize quality of life for the common good of all, for those living today and all those who come later, because the entire world is now in a permanent race to the top. Governmental corruption and incompetence would be a distant hazy memory. [77]

It sounds too good to be true. And it would be, if it was based solely on an *intuitive vision* of how things could be. But this vision is based on something completely different: an *analytical vision*. It rests on the same foundation as astronomy, medicine, physics, chemistry, and other fields of science: a collection of comprehensive principles. These principles allow the building of scientific models that, as a field matures, provide extraordinary explanatory and predictive power.

The particular branch of science we are concerned with is social system engineering. This is in its infancy when it comes to government social control models. Presently these are not engineered, but thrown together with great big lumps of intuitive insights, based on what worked and what didn't in the past. All this would change if social system engineering was a mature science.

Because the science used to create social control models is immature, the particular model that civilization uses to run itself can handle some problems and not others. The model can handle the basic needs of nations. But it cannot handle their more advanced needs, such as how to solve the sustainability problem. This places the world's social control model in the Model Crisis step of the Kuhn Cycle. Let's examine this proposition.

The Kuhn Cycle

All fields of science are built on standard accepted models of explanation and prediction. In 1962 Thomas Kuhn's *The Structure of Scientific Revolutions* stunned the scientific community with the theory that the history of these models is not a slow, progressive, evolutionary accumulation of knowledge. Instead, science is continually undergoing a predictable cycle that includes violent intellectual revolutions. This violence stems from the way old paradigms are shattered by new ones, and the way supporters of old and new paradigms battle it out until the new one wins. The five steps of the cycle are shown.

In Kuhn's terminology, a model is the shared mental model (really a problem solving process plus facts) being used by a scientific field to solve problems in that field. At first, in the **(1) Normal Science** phase, the model works so well it is supported by all. But then, as new problems arise that it cannot solve, the **(2) Model Drift** phase begins. As more and more problems remain unsolved, Model Drift increases. Eventually it deteriorates to the point that it can no longer explain and predict what it should, causing the model to enter the **(3) Model Crisis** phase. In this phase those using the model have fully awakened to the fact that their beloved model, the one that worked so well for so long, is now ready for the trash heap because it no longer works. But because they have nothing to replace it with, the model users are in crisis. They cannot make sound decisions anymore and they know it. About all they can do is try to patch and plug the old model, and use brute force to try to make it work better. While such heroic effort is commendable, it cannot be productive because the model is broken. It no longer works.

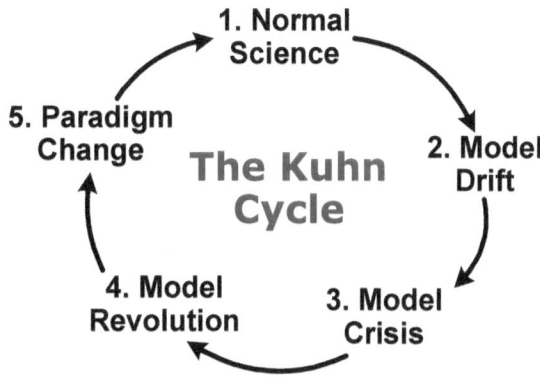

The Model Crisis phase continues until the first realistic candidate to become the new model/process appears. This initiates the **(4) Model Revolution** phase of the Kuhn cycle. In this phase the old paradigm and the candidates to become the new paradigm battle it out in a prolonged struggle for survival of the fittest. Eventually the competition evolves into a viable replacement for the old paradigm, and the jostling between those supporting the old and the new begins to quiet down. This signals the beginning of the **(5) Paradigm Change** phase, during which the new paradigm is taught to newcomers and those using the old paradigm. For major paradigms this usually takes at least a generation, because there are so many people habituated to the old paradigm that despite all evidence the new way is ten times better, they refuse to give up the old way, and take it with them to the grave.

The Paradigm Change phase is where severe change resistance occurs. People find it hard to change core beliefs. As John Kenneth Galbraith explained, "Faced with the choice between changing one's mind and proving that there is no need to do so, almost everybody gets busy on the proof."

But with the passing of enough time, the new paradigm gains the support of the majority and becomes the new **(1) Normal Science.** The cycle then starts all over again, because our knowledge about the world is never complete.

Applying the Kuhn Cycle to the Democratic Model

The fog of history hides more than we will ever know. The Kuhn Cycle allows us to peer through this fog, and see that the science of social system engineering has already passed through several cycles. Today, due to inability to solve the sustainability problem, social system engineering and the government social control models it creates are in the Model Crisis phase of the Kuhn Cycle.

In the current cycle the Normal Science is *liberal representative democracy.* (This discussion will omit the free market and corporate parts of the model for simplicity.) The current model was born in the Model Revolutions of the American and French revolutions of 1776 and 1789. Paradigm Change has taken some time. France wavered between empire, monarchy, and a democratic republic for 75 turbulent years. It has taken 200 some years for the new paradigm to displace the old one.

Today the vast majority of nations have adopted the new model. The lone large holdout, China, is still in the Model Crisis phase of an old paradigm, communism. Incremental progress is being made toward the new paradigm. It will not be long before China finds it and moves into Model Revolution, as Russia already has, and then stumbles its way through Paradigm Change, as Russia is now doing.

As good as the new Normal Science of democracy is, it could be better. The model deals awkwardly with problems like discrimination, crime, and minority interests. It has failed repeatedly with the problems of war, corruption, and economic inequality.[78] And now, for over thirty years it has failed to solve the global environmental sustainability problem, ever since it was identified in 1972 by the *Limits to Growth* project. The result is civilization is facing ecological disaster. *Thus we are now in the Model Crisis stage of the Kuhn Cycle, because the model of government currently in use is unable to solve major problems. It is no longer achieving its goal of running civilization well.*

The democratic model embodies an ambitious promise. In general the model says that if a nation's citizens are allowed to freely elect their leaders, if there are adequate checks and balances on these leaders, and if the lawful rights of citizens are protected, then the political system will behave in the best interests of the people and provide them with the best of all possible worlds, within a reasonable range since complete perfection is impossible. *By comparison to the old paradigms of dictatorship, monarchy, aristocracy, and military empire, the new Normal Science of democracy was at first a radical improvement.* But now, by comparison to what is needed, the model is broken and obsolete. It is in crisis. The present mechanisms of democracy are incapable of solving the difficult problems civilization faces today.

How then can the model be fixed? What will the new paradigm be?

The history of other branches of science holds the clues to how these questions may be answered. *In every case it was the invention of new fundamental principles and tools that allowed a new paradigm to grow on top of the old one, thus evolving the old model into the new one.*

Consider this classic example of a Kuhn Cycle: Not so long ago astronomers were unable to correctly *explain* why the heavens moved the way they did. Nor could astronomers *predict* when a comet would return. For 2,200 years, from the 6th century BC of ancient Greece up until the 16th century, the geocentric model of how the heavenly bodies moved held sway. It could explain and predict some things. But there was so much it could not that it entered the Model Crisis step. Soon Model Revolution began in earnest when Copernicus proposed in 1543 that the Earth and other planets revolved around the sun. After the telescope was invented in 1609 and was used to prove the new model was true, Paradigm Change swept the field and the heliocentric model became new Normal Science, in what has become known as the Copernican Revolution. Today the heliocentric model, Kepler's three laws of planetary motion, and Newton's law of universal gravitation and his three laws of motion form the foundation of all of astronomy.

Notice how the new paradigm grew on top of the old one, by discovery of a series of powerful new principles and tools. The first breakthrough was invention of the heliocentric model. This kept the idea of heavenly bodies moving around a central single body and rejected the rest of the geocentric model. But the new model lay unaccepted for decades, until the critical mass of further inventions necessary for model maturity appeared. These included the telescope, Kepler's three laws of planetary motion, and Newton's law of universal gravitation and his three laws of motion. Once these inventions were applied, the new model's explanatory and predictive power was so much better than the old one it became the new paradigm, and another Kuhn Cycle was complete.

This shows that *for an old model to progress to a new one, a number of breakthrough discoveries are required*, enough to achieve a new critical mass of explanatory and predictive power. The previous chapter explored what these discoveries might be. If they can be accelerated, then a field can move from Model Crisis to Model Revolution to Paradigm Change and finally to the new Normal Science in as little as half a generation, though unfortunately it usually takes several.

The Critical Mass Components
of the New Model of Social System Engineering

Like the Copernican Revolution, the new model will evolve from the old one, using a critical mass of newly invented components. Revisiting the previous chapter, here are the five components that appear to be needed:

Component 1 – A System Modeling Tool

The Copernican Revolution ushered in the greatest change the field of astronomy has ever seen, with its radical notion that the sun, and not the earth, was the center of our little nook in the universe. It would help greatly if the first new component is as staggeringly powerful as the one that launched the Copernican Revolution.

As this book and others such as *Limits to Growth* have demonstrated, a new component that may prove to be just as crucial already exists: system dynamics. This tool reveals the structure of social systems just as clearly as the Copernican model showed the true structure of the heavens. There are other modeling tools that will be also necessary, but system dynamics has the advantage of simplicity and emphasis of feedback loops.

However, like the invention of the heliocentric model, system dynamics alone is not enough to achieve the critical mass necessary. More new components are needed.

Component 2 – The Boundaries of Memetics

Another key component of the new model appeared in 1976 in *The Selfish Gene*, a book by Richard Dawkins, a British evolutionary biologist. In the final chapter Dawkins dropped an intellectual bombshell when he wrote the following electrifying words. In so doing he coined a new word that has now entered the Oxford Dictionary: (Italics are his)

"For an understanding of the evolution of modern man, we must begin by throwing out the gene as the sole basis of our ideas on evolution.

"What after all, is so special about genes? The answer is they are replicators. All life evolves by the differential survival of replicating entities. The gene, the DNA molecule, happens to be the replicating entity that prevails on our own planet. There may be others. If there are, provided certain other conditions are met, they will almost inevitably tend to become the basis for an evolutionary process.

"But do we have to go to distant worlds to find other kinds of replicator and other, consequent, kinds of evolution? I think that a new kind of replicator has recently emerged on this very planet. It is staring us in the face. It is still in its infancy, still drifting clumsily about in its primeval soup, but already it is achieving evolutionary change at a rate that leaves the old gene far behind.

"We need a name for the new replicator, a noun that conveys the idea of a unit of cultural transmission, or a unit of *imitation*. 'Mimeme' comes from a suitable Greek root, but I want a monosyllable that sounds a bit like 'gene.' I hope my classicist friends will forgive me if I abbreviate mimeme to *meme*. If it is any consolation, it could alternatively be thought of as being related to 'memory,' or the French word *même*. It should be pronounced to rhyme with 'cream.'

"Examples of memes are tunes, ideas, catch-phrases, clothes fashions, ways of making pots or of building arches. Just as genes propagate themselves in the gene pool by leaping from body to body via sperm and eggs, so memes propagate themselves in the meme pool by leaping from brain to brain via a process which, in the broad sense, can be called imitation. If a scientist hears, or reads about, a good idea, he passes it on to his colleagues and students. He mentions it in his articles and his lectures. If the idea catches on, it can be said to propagate itself, spreading from brain to brain."

*A **meme** is a copied mental instruction capable of affecting behavior.* All memes are learned from others, either directly from other people or indirectly through a transmission medium, such as books or television. All words, unless you made one up yourself, are memes. All learned values, such as "trustworthiness is good," are memes. Reading, writing, and arithmetic, because we learned them from others, are gigantic sets of interrelated memes. Thus the entire foundation of all fields of traditional knowledge, such as biology, physics, and mathematics, are memes.

Memes are a concept so intriguing and vitalizing they breath a flurry of insights into any discussion of how social systems work, just as the discovery of gravity did for astronomers. Memes, combined with the three steps of evolution, instantly provide a comprehensive explanatory foundation for all of human learning, culture, and cultural evolution. This is no small feat.

All of culture is memetic, including every last word in a constitution. Thus if we can understand how and why memes serve to drive social control models, we can understand how and why models of government work. Once we understand that, we will be a giant step closer to being able to *proactively* engineer social control models, instead of letting them evolve *reactively*, as we do now.

Without memes we would have been unable to build the Dueling Loops model, because at the heart of the model is the creation and transmission of memes by those seeking supporters.

Memeticists have now had 30 years to mature their field. It has not advanced far, as many who were attracted to it fervently hoped. But the concept of memes has been chewed on so thoroughly that I had no trouble finding numerous interpretations and elaborations that greatly helped me in my work.

The field's boundaries are complete, however, as this definition from the Journal of Memetics shows:

> "**Memetics** is the theoretical and empirical science that studies the replication, spread and evolution of memes. Its core idea is that memes differ in their degree of 'fitness', i.e. adaptation to the socio-cultural environment in which they propagate. Because of natural selection, fitter memes will be more successful in being communicated, 'infecting' a larger number of individuals and/or surviving for a longer time within the population. Memetics tries to understand what characterizes fit memes, and how they affect individuals, organizations, cultures and society at large." [79]

Component 3 – The Fundamental Principles of Memetics

Although the boundaries of memetics are known, the field itself remains immature because its fundamental principles are incomplete. The most important principle of all—that memes behave as evolutionary replicators just as genes do—is well established. The rest remain undiscovered. Because they hold the key to understanding the social side of the human system, and that is arguably more important than the technical side as we have now so suddenly discovered, *the most important frontier in all of science has yet to be explored.*

Only a few intrepid innovators have set foot in this new science. Almost no one has followed them.

The urgent need to find the fundamental principles, once widely known, should serve to attract the first wave of explorers. It is they who will soon provide the foundation for solving the toughest and most important problem our species has ever faced: How can global civilization, which depends on the coordinated behavior of all seven billion of its members, govern itself effectively enough to avoid mass ecocide?

The problem will only be one percent solved once the fundamental principles of memetics are known. The rest, as Thomas Edison knew, is the ninety nine percent perspiration needed to DO something with this knowledge. As far as we can tell, the bulk of the work will center on the next two components.

No Predictions Here

You may be wondering why this chapter has not provided any detail about what a permanent race to the top would look like. We leave that to the futurists.

The important thing is to engineer structural incentives into the system that cause the system's dominant agents to automatically evolve the system toward an optimum future. What that will be exactly, particularly more than one generation from now, no one knows. But with proper analysis and design, we do know that it will be desirable, equitable, and about the best that a species endowed with the ability of hyper-reasoning can achieve.

And it will be sustainable.

Component 4 and 5 – Memetic Calibration Techniques and Fundamental Social Control Model Parts

Both of these areas are nearly completely unexplored. For example, I have yet to run across even the simplest memetic driven simulation model that has been calibrated, though there must be some out there. This includes my own. Calibration takes a lot of expertise, time, and money for the results to be statistically valid. As to whether any fundamental social control model parts exist in system dynamics form, they may.

This review of what it will take to achieve critical mass has been necessarily speculative. More inventions will be needed. Once there are enough for engineers to design solid, flexible, self-managing human systems, we will have the real breakthrough: the ability to proactively engineer large social systems so that the systems achieve their design goals.

This has never been done. All past social control models, including cities, corporations, nations, and political parties, evolved through trial and error and intuition. They were never rigorously engineered. They couldn't be, because the means did not exist.

But once it does, there will be little stopping progressives from:

Making the Race to the Top Permanent

Page 169 lists the ten steps of the System Improvement Process. *To make the race to the top among politicians as permanent as permanent can be, we must successfully execute all ten steps in the process.*

The full analysis results are summarized on the next page. Two proper coupling subproblems were found, giving a total of four subproblems. HLPs are high leverage points. Here the System Understanding step is called Analysis.

The Analysis contains strong, promising hypotheses of the root causes and high leverage points. It is up to those who take up the challenge presented in this work to prove or disprove these hypotheses, and if disproved, to find the correct root causes and HLPs.

The Solution Convergence row contains well thought out solution elements designed to push on the identified high leverage points. These elements are ready for experimentation. If they prove to resolve the root causes, then with refinement and further experimentation they can be scaled up to solve the subproblems. The scaling up is where Implementation occurs. Because of a long smooth scaling up via progressively larger experiments, there is no sudden big bang, which too often results in a big bust. Nor is there a wild, intuitive guess as to whether a solution will work. Instead, we have solution elements that are the output of a process that fits the problem, rigorous analysis, and experimentation. Solutions like these are likely to work the first time.

Summary of Analysis Results of Executing SIP on the Global Environmental Sustainability Problem

1. Problem Definition		How to achieve global environmental sustainability in terms of the desired system goal state			
	Subproblems	**A.** **How to Overcome Change Resistance**	**B.** **How to Achieve Life Form Proper Coupling**	**C.** **How to Avoid Excessive Model Drift**	**D.** **How to Achieve Environmental Proper Coupling**
2. Analysis — A. Find immediate cause loops	Subproblem symptoms	Successful opposition to passing proposed laws for solving the problem	Large for-profit corporations are dominating political decision making destructively	Failure to correct failing solutions when they first start failing	The economic system is causing unsustainable environmental impact
	Improperly coupled systems	Not applicable	Corporate and human life forms	Not applicable	Economic and environment systems
	Analysis model	Basic Dueling Loops of the Political Powerplace	Complete Dueling Loops model. This adds the Alignment Growth loop.		The World's Property Management System
	Immediate cause dominant loops	The Race to the Bottom among Politicians		Intelligent Adaptation loop in evolutionary algorithm model	Industrial Growth and Limits to Growth (the IPAT factors)
B. Find inter. causes, LLPs, SSs	Intermediate causes	The universal fallacious paradigm, primarily Growth Is Good	Disagreement from corporate proxies on what to do	Laws giving corporations advantages over people	Externalized costs of environmental impact
	Low leverage points	More of the truth: identify it, promote it, magnify it	Logical and emotional appeals and bargaining	Trying to directly reverse laws that favor corporations	Internalize costs
	Symptomatic solutions	Technical research, environmental magazines and articles, awareness campaigns, marches, sit-ins, lawsuits, lobbying, etc.	Corporate social responsibility appeals, green investment funds, NGO/corporate alliances, etc.	Media use, campaigns, lobbying to get old laws repealed	Main solutions at system level: regulations and market-based, like pollution taxes and tradable permits. At agent level main solutions are 3 Rs and collective mgt.
	C. Root causes of intermediate causes	High political deception effectiveness	Mutually exclusive goals between top two social life forms, *Corporatis profitis* & *Homo sapiens*	Low quality of political decisions	High transaction costs for managing common property sustainably
	D. Loops that should be dominant to resolve root cause	You Can't Fool All of the People All of the Time	Alignment Growth		Sustainability Growth and Impact Reduction
	E. High leverage point to make those loops go dominant	General ability to detect political deception	Correctness of goals for artificial life forms	Maturity of the political decision making process	Allow firms to appear to lower transaction costs
3. Solution Convergence		Nine solution elements	Corporation 2.0, *Corporatis publicus*	Politician Decision Ratings	Common Property Rights
4. Implementation		Not yet ready for implementation because process execution is incomplete.			

Due to depth of analysis, the root causes and HLPs of the first three sub-problems are *generic*. Therefore so is their solution. The fourth subproblem is the environmental sustainability problem. Thus the solution to the first three subproblems applies not just to environmental sustainability, but to all problems whose solution would benefit the common good. This is required if we are to achieve the full potential of a permanent race to the top.

This book argues that the change resistance subproblem must be solved before it will be possible to implement a full solution any other subproblem. *Change resistance is thus the real problem to solve.*

But that's not the way the world sees it. Instead, most problem solvers have a distorted view of reality, as the diagram below shows. Each block contains the four subproblems from the Summary of Analysis.

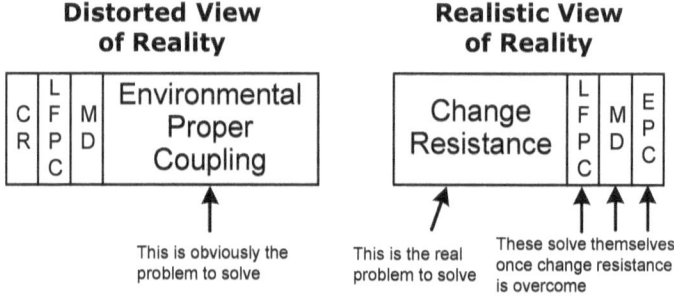

The distorted view arises from the way most problem solvers approach the sustainability problem. They use common sense and the same methods that work on normal everyday problems. Thus environmental proper coupling is obviously the problem to solve. This mindset makes the change resistance (CR), life form proper coupling (LFPC), and model drift (MD) subproblems small or invisible. But a realistic view of reality sees a different picture. Change resistance is big because it is *the* problem to solve. Environmental proper coupling is actually small and insignificant, because it will solve itself once change resistance is overcome.

All conscious decisions are the based on mental models. The distorted view is an example of a flawed mental model of a problem. This is common. To illustrate how common, on the next page is an actual drawing from a consulting case at the Organizational Learning Center at MIT.[80] A company's managers were having trouble reducing total time from customer order to acceptance. The managers viewed order fulfillment lead time as the biggest delay. It was thus seen as the real problem to solve, even though the managers had data showing otherwise. Their mental model was so distorted that they drew *a scale model* reflecting their erroneous thinking. Take a look at their amazing blooper:

Notice how the order fulfillment block is the biggest. But it should be the smallest! This is the same trap those working on the sustainability prob-lem have fallen into. Once

Current Supply Chain Cycle Time

Goal: cut cycle time by 50%, from 182 to 91 days.

Manufacturing Lead Time	Order Fulfillment Lead Time	Customer Acceptance Lead Time
75 Days	22 Days	85 Days

◄———————— 182 Days ————————►

their mental models were set, they never changed. The moral of this story is to be skeptical about everything, particularly your own mental models of the world and the processes you are using to solve problems.

The goal of the System Improvement Process is to make the entire process of solving difficult social problems as efficient, effective, and as fast as possi-ble. If you think your way through the Summary of Analysis and what each of its cells means, turning back to the process steps on page 169 as needed, you will have the beginning of a vision of how the *complete* process can be exe-cuted for the *complete* sustainability problem.

Once you see that, then you may conclude, as I have, that there really is a path forward for progressives to achieve their ultimate goal—permanently.

How Progressives Can Find Their Way Again

This book promised to crack the mystery of why progressives are stymied and how they can find their way again. Here is a look back at how we unrav-eled that mystery, and a brief look at one way to move forward:

The progressive movement finds itself in the predicament of being blocked from achieving its ideals. This is an eternal paradox, because progres-sives throughout history have always tried to optimize the human system for the good of all, while the opposition has done just the opposite: optimization of the system for the good of the few. It would seem that that now that democ-racy is the norm, the system should welcome such unselfish effort. But no. It is the selfish side that is winning. This causes the system as a whole to lose.

The *intermediate cause* of this paradox is systemic change resistance. When change resistance is present a problem is several orders of magnitude more difficult to solve, because vast portions of the system's behavior as an emergent whole must be changed. Most progressive problem solving failures are due to strong and/or system wide change resistance, while most successes are due to the fact that change resistance was weak and/or local. *Few activists can tell the difference.* The result is success on a few problems, failure on the rest, and tremendous frustration.

This frustration should end, now that we know the *root cause* of change resistance is a dominant race to the bottom in a social structure called The Dueling Loops of the Political Powerplace. This provides a satisfying hypothesis for the main reason progressives are stymied. It explains why they are unable to solve their top problems and why corruption is so common, despite repeated efforts at reform. It also shows The Battle for Niche Succession is underway. *Homo sapiens* is losing badly to the New Dominant Life Form, which is the modern corporation and its allies, notably the rich.

This hypothesis demonstrates the critical importance of being able to "see" the structural behavior of social problems, such as the way the models in this book let us see much more clearly how political powerplaces work. *A central message of this book is that until activists can see the social structure of the problems they are attempting to solve, they will be unable to tell the high leverage points from the low leverage ones, and will be unable to solve difficult problems reliably.* Only seeing social structure allows correct diagnosis, and only a correct diagnosis allows the patient to be correctly treated.

For example, only after they could "see" and understand the structure of molecules could chemists reliably solve their problems. The same pattern holds for physicists, astronomers, biologists, doctors, architects, and more. Until they could correctly comprehend the structure of what they were working on, they were blind and groped around in the dark for centuries, often with disastrous results and always with slow progress.

Progressives tend to be a minority force compared to the status quo. They must push on high leverage points because they simply do not have the force needed to make pushing on low leverage point work. Therefore a structural analysis with a formal model is required.

This is a bit of work. But once the structure of the fundamental behavior of the problem becomes visible, where the high leverage points are is relatively obvious. *Each high leverage point is a solution strategy.* There are many ways to push on high leverage points, which is the same thing as saying there are many ways to implement a strategy. To illustrate how this can realistically be done, this book presented seven solution elements: Freedom from Falsehood, the Truth Test, Truth Ratings, Corruption Ratings, No Servant Secrets, the Sustainability Index and Decision Ratings. These are only educational examples, however. Much further analysis, experimentation and iteration remains.

Progressive philosophy was carefully defined as a comprehensive rationale and value set whose goal is optimizing the human system for the common good of all and their descendents. Thus progressives are humanists. *Therefore, if the Dueling Loops exist, then progressives may not realize it, but their central strategy is the high road of winning the race to the top.*

This has been their strategy all along. The only thing that has changed is it now has a name.

Notice, for example, how virtuous politicians hesitate to resort to *ad hominem* attacks as an election draws near, while their degenerate opponents go on the attack early and often. Notice how progressive writers and think tanks stick to the truth, while those serving the New Dominant Life Form mix fallacies with fact routinely. And so on. These behavior traits are the consequence of relying on race to the top or bottom strategies.

The Dueling Loops also explain why one side is so dependent on a dogmatic ideology, while the other side has no dogma and is more of a flexible philosophy. Pursuit of the truth allows a loose, multifaceted approach based on an evolving, constantly challenged philosophy. But reliance on falsehood requires a tight, dogmatic, centralized managerial approach, both to manufacture the lies and market them. The aggressive, well orchestrated marketing of these lies requires those peddling them to dogmatically stay "on message," and thus appear to be more consistent and hence more true. This prevents the web in "Oh what dangerous webs we weave, when we practice to deceive" from unraveling.

All this leads to how progressives can find their way again:

Step 1 – *Progressives need to verify the Dueling Loops exist, and then agree that their top solution strategy needs to be making the race to the top go dominant.* Or if the Dueling Loops do not exist, then they need to find the real root cause of the paradox and develop an alternate solution strategy. Either way, the strategy becomes their explicit new goal.

Step 2 – The second step follows logically. *The only reliable way to achieve a difficult goal is to develop a plan and then implement the plan.* In this case the best way to do this is to adopt a process that fits the problem, just as scientists and business managers do. A *strategic process* that fits the problem is Analytical Activism, which is the use of the Analytical Method to achieve activist objectives. However, any process that fits the problem will do.

The **Analytical Method** is a nine step generic process combining the power of the Scientific Method with the use of formal process to solve any type of problem. Step two requires selection of a *tactical process* fitting your particular problem. An example of one that fits most difficult activist problems is the System Improvement Process.

How to apply these processes to activist problems is described at length in the larger companion to this book: *Common Property Rights: A Process Driven Approach to Solving the Complete Sustainability Problem.* This gives a thorough introduction to solving difficult activist problems using the most efficient and effective methods available. The book practices **Analytical Activism,** which is a problem solving approach allowing activists to base their key decisions on sound reasoning and facts, instead of intuition and optimism.

The Analytical Method

1. Identify the problem to solve.
2. Choose an appropriate process.
3. Use the process to hypothesize analysis or solution elements.
4. Design an experiment(s) to test the hypothesis.
5. Perform the experiment(s).
6. Accept, reject, or modify the hypothesis.
7. Repeat steps 3, 4, 5, and 6 until the hypothesis is accepted.
8. Implement the solution.
9. Continuously improve the process as opportunities arise.

Step 3 – Let's assume the Dueling Loops exist. The third step, once the second step is well underway and begins to succeed, is to *raise the bar and make the goal a permanent race to the top.* This becomes the core of the next generation model of democracy. Until permanent dominance of the race to the top loop is achieved, progressives will find themselves struggling to solve a never ending series of waves of activist problems, due to the cyclic nature of the Dueling Loops.

What would a permanent race to the top look like? What will happen when politicians are in a permanent state of constructively competing to see who can do the best job of optimizing the human system for the common good of all and their descendents? Will they employ Decision Ratings to radically improve the output of political systems or will they find something even better? What will a global society with no corruption and no control by special interests be like? Where will this extraordinary mode change take civilization over the next few centuries?

No one knows, because the race to the top has never stayed dominant for long. But we do know that compared to its predecessor models of ruthless dictatorship, oppressive monarchy, let-them-eat-cake aristocracy, and destructive military empire, the present model of democracy is an improvement by an order of magnitude over the old ones.

The next model will make just as large a leap.

Appendix

The mission of Thwink.org is to help solve the global environmental sustainability problem using the most efficient and effective methods available. To do this we have created a variety of tools and educational materials. Here are some that relate to this book:

The Dueling Loops Videos

A picture is worth a thousand words. If you would like to learn more about the Dueling Loops, watch the Dueling Loops video series. Each video runs 5 to 10 minutes. To find them, enter "dueling loops videos" in the search box.

The videos include some material not in the book. A highlight is Part Two, which extracts the Competitive Spiral from Jared Diamond's *Collapse*, then adds the Cooperative Spiral and the other two loops, arriving at the model shown above. This is then rearranged into the basic Dueling Loops shape, thus illustrating the timeless ubiquity of the shape.

The strategic purpose of social system engineering is to allow civilization to stay in the Cooperative Spiral, which is also the race to the top.

The Progressive Paradox Film

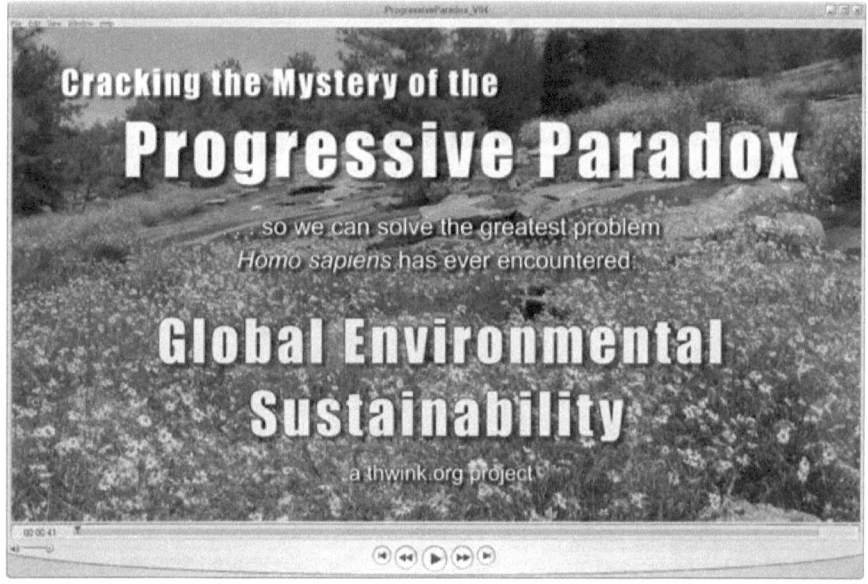

Thwink.org has completed a downloadable two hour high definition film. This presents additional material not covered in this book. To find it on the sprawling Thwink.org site, enter "paradox film" in the search box.

The highlight of the film is introduced near the end: how the System Improvement Process works.

We can crack the mystery of the Progressive Paradox by completing enough cells in the process grid to make the race to the top go dominant. This requires *only four out of ten cells*: problem definition plus the three cells in the change resistance column. At this point change resistance to solution adoption is overcome, and the system "wants" to achieve proper coupling, which is the next column in the grid.

If we can complete *all ten cells*, then it becomes possible for that state to become permanent. Otherwise it appears that due to the cyclic nature of the Dueling Loops the solution will degrade and problem will recur.

Endnotes

[1] "…and from changing the politicians to changing the system." This paraphrases what Barack Obama wrote in an editorial on January 4, 2007 in the Washington Post:

> **"A Chance To Change The Game** – This past Election Day, the American people sent a clear message to Washington: Clean up your act.
>
> "After a year in which too many scandals revealed the influence *special interests* wield over Washington, it's no surprise that so many incumbents were defeated and that polls said '*corruption*' was the grievance cited most frequently by the voters.
>
> "It would be a mistake, however, to conclude that this message was intended for only one party or politician. The votes hadn't even been counted in November before we heard reports that *corporations* were already recruiting lobbyists with Democratic connections to carry their water in the next Congress.
>
> *"That's why it's not enough to just change the players. We have to change the game."*

Obama intuitively senses something like the Dueling Loops are hard at work. "The influence *special interests* have over Washington" and "*corruption* was the grievance cited most frequently by the voters" are telltale symptoms the race to the bottom is dominant.

A key principle of systems thinking is that solutions based on treating the symptoms of a problem don't work because they do not resolve the underlying cause. A corrupt politician is a symptom of a broken system. Changing the politician by electing a better one usually fails in the long run because the same underlying forces are at play.

See: www.washingtonpost.com/wp-dyn/content/article/2007/01/03/AR2007010301620.html

[2] "The right process will produce the right results" is the title of section two of The 14 Principles of The Toyota Way. See http://www.si.umich.edu/ICOS/Liker04.pdf for a seven page summary of these principles. For much more, get *The Toyota Way* book they are from. An example of The Toyota Way is *"It costs less to build it right the first time than to fix it later."* This is one thing the System Improvement Process attempts to do, with its emphasis on getting the analysis right in step 2 so step 3 is correct and step 4 works, *the first time.*

[3] "Therefore the social side is the crux of the problem and must be solved first." For additional research on this insight see our 2010 paper on *Change Resistance as the Crux of the Environmental Sustainability Problem.*

[4] "What they should be working on instead is how to get the horse to decide to drink." When I showed my editor-in-chief (my wife) this paragraph, she immediately said, "Oh that's easy. You just make the horse thirsty. Run him around a little and he will get thirsty." If only the sustainability problem was as easy to solve.

[5] The material on the superficial definition of progressivism is paraphrased from the third definition at en.wikipedia.org/wiki/Progressivism on June 26, 2007, plus the addition of peace.

[6] Optimizing the human system for the good of all tends to lead to the main sub goal of maximizing quality of life instead of quantity of wealth. In a world of finite resources, this greatly reduces pressures on environmental limits, because maximization of quality of life does not consume anywhere near the resources that maximization of quantity of wealth does. A focus on quality of life also makes solving problems like poverty and excessive inequality of wealth easier. Thus another definition of **progressive philosophy** is a comprehensive rationale and value set whose goal is maximizing quality of life for the good of all.

[7] The first definition of degenerate is from dictionary.com, unabridged, version 1.1, Random House Inc, retrieved May 27, 2007. The second definition is from *Choose the Right Word*, by Senator Hayakawa, 1994, page 109.

[8] Quote from www.sustainabilityinstitute.org/pubs/Leverage_Points.pdf, page 1. The paper does have some nice insights on classes of leverage points. However, defining leverage points as "places within a complex system where a small shift in one thing can produce big changes in everything" fails to define what a "small shift" is. For example, all it takes to shift a country to support of the Kyoto Protocol treaty is the right signature on the right document. But what about all the effort it takes to persuade the person who sits down and pens that signature?

Thus any definition of leverage points must include the total input effort required to make the change. But this is not seen in the paper. For example, it says "But if there is a delay in your system that can be changed, changing it can have big effects." It's as if it doesn't matter at all how much effort it will take to change the delay. All that seems to matter is that if it is changed, that can have big effects.

[9] Meadows, D. H. & Meadows, D. L. & Jorgen, R. & Randers J. 1972. *The Limits to Growth*. Potomac Associates. Page 24.

[10] World Commission on Environment and Development, 1987. Our Common Future. Oxford University Press. The quote is from the back cover.

[11] Constanza, Cumberland, Daly, Goodland and Noregaard. 1997. *An Introduction to Ecological Economics*. St. Lucie Press. Pages 206-207.

[12] IPCC. 2007. Fourth Assessment Report. Climate Change 2007: Synthesis Report. Summary for Policymakers. www.ipcc.ch/pdf/assessment-report/ ar4/syr/ar4_syr_spm.pdf. Page 18.

[13] The "squandered the last 30 years" quote is from the third edition of *Limits to Growth*, 2004, pages xiii and xvi .

[14] Randers, Jorgen. 2000. *Limits to growth to sustainable development or sustainable development in a system dynamics perspective*. System Dynamics Review, vol 16 No 3 page 223.

[15] Senge, Peter. *The Fifth Discipline: The Art & Practice of the Learning Organization*, 1990. Currency Doubleday. Page 88.

[16] Cunningham, William et al. 1998. *Environmental Encyclopedia*. Page 1055.

[17] United Nations Division for Sustainable Development, http://www.un.org/esa/sustdev/documents/agenda21/index.htm July 9, 2008.

[18] Global Footprint Network. 2007. *Ecological Footprint Overview.* www.footprintnetwork.org/gfn_sub.php?content=national_footprints. De-cember 20, 2007. Note that the ecological footprint only measures pollution and renewable resource use. It does not include non-renewable resource depletion. Thus getting the world's footprint down to the one planet line is only half the battle. The other half, and probably the harder half, is getting the non-renewable resource use rate down to zero. Gulp.

[19] WikiPedia, http://en.wikipedia.org/wiki/Ecological_footprint#Ecological_footprint_studies_in_the_United_Kingdom July 9, 2008.

[20] World Wildlife Fund, 2006, *Annual Report.* The Ecological Footprint graph is also in the third edition of *Limits to Growth,* 2004, which is where I first encountered it. The graph has been redrawn and the dots added.

[21] Source of milk allergy data: www.mayoclinic.com/health/milk-allergy/DS01008.

[22] Orwell, George. 1946. *Politics and the English Language.* The quote is on the last page.

[23] The quote about Václav Havel is from cestazmeny.net/veracity-in-politics.html. Havel was famous for his essays, most particularly for his brilliant articulation of "Post-Totalitarianism," a term used to describe the modern social and political order that enabled people to "live within a lie". (This sentence is from the Wikipedia entry on Václav Havel.)

[24] The quote on fear is by George Gerbner, past dean emeritus of the University of Pennsylvania's Annenberg School for Communications, from an obituary in the Washington Post on January 2, 2006, at www.washingtonpost.com/wp-dyn/content/article/2006/01/02/AR2006010200577.html.

[25] The Moynihan quote is from www.cnn.com/SPECIALS/cold.war/episodes/06/book. Further material is from www.cnn.com/books/reviews/9810/22/secrecy.cnn and www.encyclopedia.com/doc/1G1-53972635.html.

[26] The Frank Rich book review is from www.nytimes.com/2006/09/22/books/22book.html.

[27] Whether these really are the five main types of political deception is an educated guess. Their purpose is to help initiate the decomposition of large scale political deception into useful classifications, such as the way there are six types of quarks: up, down, bottom, top, strange, and charmed. Deception classification will probably not begin in earnest until more details are known about the specific system dynamics structures that different deception meme types employ to achieve replication success.

[28] "They cannot tell a bigger truth" holds for the Boolean sense that something is either true or false. This allows the model to be a useful simplification of reality. Actually a bigger truth is possible if the quality of a statement is considered. For example, there may be many ways to balance the budget. Some will solve many problems, some will solve only a few, and some will cause more problems than they solve. Modeling this finer shade of behavior was not needed to analyze the particular problem under consideration.

[29] The actual quote is "All power tends to corrupt and absolute power corrupts absolutely." Lord Acton. 1887. Letter to Mandell Creighton. From en.wikiquote.org/wiki/Lord_Acton.

[30] The phrase New Dominant Life Form is designed to emphasize the way the modern corporation and its allies are a true life form and are currently *the* dominant one on the planet. Numerous authors have written about the power and damaging behavior of corporations. Some of the books I've read are *Gangs of America: The Rise of Corporate Power and the Disabling of Democracy,* by Ted Nace, 2003; *When Corporations Rule the World,* by David Korten, 2001; and two by Sharon Beder: *Global Spin: The Corporate Assault on Environmentalism,* 2002, and *Suiting Themselves: How Corporations Drive the Global Agenda,* 2006.

[31] The table of the world's 100 largest economies is from *Global Inc.: An Atlas of the Multinational Corporations,* 2003, by Gabel and Bruner.

[32] Despite my best efforts, many readers of drafts of this book concluded it paints corporations as a demonic enemy that must be vanquished to solve the problem, when it actually doesn't do that at all. For example, one reader wrote that the *"anti-corporation [language] has to be scrubbed to [become a] synergistic partnership everywhere."* Another test reader wrote *"Also, on the Corporations piece, your language demonizes them. No question about it. Whether it's your intent or not, the impact is off-putting (spoken like a former corporate lackey). I think I might have mentioned a saying in my field to you previously: 'Take a good person and put them in a bad system, and the system wins every time.' "*

The reasoning seems to be that since I'm saying that corporations have done a bad thing by behaving unsustainably, then I'm saying corporations are bad and we environmentalists and progressives should feel hostile toward them. This is an unjustified conclusion. What I'm trying to say is that we need to find the underlying causes of undesirable agent behavior and fix that, by changing the system.

To minimize reactions like the above, the section on the fundamental attribution error and the box on The Corporation DVD were added. Note how the second reader is aware of the fundamental attribution error, as shown in their last sentence. The first reader realizes that even though corporations are part of the problem, they can be part of the solution.

The reasoning seems to be that since I'm saying that corporations have done a bad thing by behaving unsustainably, then I'm saying corporations are bad and we environmentalists and progressives should feel hostile toward them. This is an unjustified conclusion. What I'm trying to say is that we need to find the underlying causes of undesirable agent behavior and fix that, by changing the system.

[33] The passage about the fundamental attribution error is from the best book available on system dynamics: *Business Dynamics: Systems Thinking and Modeling for a Complex World,* by John Sterman, 2000, page 28.

[34] "Place a good person in a bad system, and the system wins every time." This quote was provided by Peter Hess in a private email on June 14, 2007. He attributed it to his previous field of human resource management.

[35] The image of *The Corporation* DVD was downloaded on June 14, 2007 from www.zeitgeistfilms.com/videocatalog/images/Corporation.DVD.jpg.

[36] The key provision quote is from *When Corporations Rule the World*, by David Korten, 2001, page 167.

[37] The power of the WTO quote is from *Suiting Themselves: How Corporations Drive the Global Agenda*, by Sharon Beder, 2006, page 118.

[38] Experimentation is just beginning. On February 16, 2007 Michael Hollcraft ran The First Experiment (see the Thwink.org website for details) for the very first time on a group of 17 people. This is a randomized, controlled, double blind experiment to test the hypothesis that *even a very brief exposure to the Truth Test can raise a person's ability to detect political deception.* The results showed the treatment group (the one exposed to the Truth Test) voted for politicians employing deception 44% less often than the control group (the one not exposed to the Truth Test). This supports the hypothesis. However, group size was too small for the results to be statistically valid at the 95% confidence level. We plan to run this experiment again on larger groups, as well as design and run many other experiments. If you would like to run the experiment, please go to the website and do a search on "First Experiment."

[39] The quote about reactions to Jay Forrester's work is from "The Beginning of System Dynamics" at sysdyn.clexchange.org/sdep/papers/D-4165-1.pdf. This was a "Banquet Talk at the international meeting of the System Dynamics Society, in Stuttgart, Germany, July 13, 1989."

[40] The story of the articulate man from Harlem is from "The Beginning of System Dynamics" in the reference above.

[41] The most interesting account of the complete problematique I've seen is at www.cwaltd.com/pdf/clubrome.pdf.

[42] However simple these solutions may appear today, they were actually complex solutions to complex problems. The reason these solutions appear simple today is the components involved are now taken for granted. For example, we may see democracy as very simple—you just let people elect their leaders. But that requires an independent judiciary to enforce the laws required to do that, various checks and balances so that no one elected body or official can abuse there power, and so on. A democracy cannot be defined in less than the length of a constitution. Thus the concept of democracy is simple, but the actual solution is not.

[43] The image is from tsa.ucsf.edu/~snlrc/encyclopadia_romana/greece/hetairai/diogenes.html.

[44] The article about Julian Burnside is from The Age, an Australian newspaper, at: www.theage.com.au/news/national/pollie-graph-idea-to-stamp-out-porkies/2007/05/14/1178995076746.html.

[45] The article with the quote on "truth predictor software" is from www.ft.com/cms/s/06adcbce-5345-11db-99c5-0000779e2340.html.

204 The Dueling Loops of the Political Powerplace

[46] The quote on "free communication of thoughts and ideas" is from: www.ambafrance-uk.org/Freedom-of-speech-in-the-French.html.

[47] Source of testimony on corporate bond ratings: hsgac.senate.gov/032002lieberman.htm.

[48] Source of Humanity's Ecological Footprint graph: World Wildlife Fund, 2006 Annual Report with improvements. The graph is also in the third edition of *Limits to Growth*, 2004, which is where I first encountered it.

A special note: There is an important limitation of the Ecological Footprint: it considers only *renewable resources use rates* and *pollution rates*. Its calculation does not include *nonrenewable resource depletion rates*. For the system to be sustainable, all three rates must be sustainable. So think of the Ecological Footprint as one of society's first steps in its search for a mature sustainability index, and keep in mind it does not reflect *nonrenewable resource depletion rates*.

[49] The World and USA data is from *Ecological Footprint of Nations 2005 Update*, at http://www.ecologicalfootprint.org/pdf/Footprint%20of%20Nations%202005.pdf on June 9, 2007. For the world, 21.91 / 15.71 = 139%. For the USA 108.95 / 20.37 = 535%.

[50] The next model iteration will probably change the Repulsion to Corruption range to vary from zero to 100%, which is much easier to work with.

[51] Regarding "The first link in an auto-activation chain must be activated manually." – This is done by a carefully engineered *precipitating event*. For more on this please see the *Analytical Activism* book.

[52] Source of quotes on niche: *Environmental Encyclopedia*, by Cunningham et al, Second Edition, 1998, page 703.

[53] Source: *Ecology Instant Notes*, by Mackenzie, Ball, and Virdee, 2001, page 18.

[54] Source of species information: wikipedia.com, starting at Human Evolution. It has since been proposed that *Homo floresiensis* may be not be a new species after all, but an occasional mutant thought to occur in one out of every 500 to 2,000 births.

[55] Source of quote: trc.ucdavis.edu/catoft/EVE101/Lec8c1.htm.

[56] Source of graph: www.geo.arizona.edu/Antevs/nats104/00lect19comptn.gif.

[57] The definition of humanism is from en.wikipedia.org/wiki/Humanism on November 17, 2006.

[58] Source: *Leos Strauss and the Neo-Cons at War*, www.logosjournal.com/mason.htm.

[59] Ibid.

[60] Source: www.disinfopedia.org/wiki.phtml?title=Leo_Strauss.

[61] The principle of naming processes this way is suggested in *The Six Sigma Handbook*, by Thomas Pyzdek, 2003, page 253. The original idea is from Hammer and Champy, 1993, page 118. Further examples are:
- Product development – The concept to prototype process
- Order fulfillment – The order to payment process

• Service – The inquiry to resolution process

Note the process could be modeled using system dynamics. This would allow deep design analysis, by examining various design and investment scenarios. It would also allow the model to be calibrated and maintained, so that continuous process improvement could be done efficiently, effectively, and proactively.

How many governments have a comprehensive process map? How many of these are using it to formally continuously improve the process, until it is so good that it automatically allows achieving the greatest common good for all? When it comes to quality of political decision making, who are the most important people: the politicians or the process managers?

[62] Later even a 5% favoritism rating will be too high, as structures are built that cause a zero tolerance to corruption. This will cause favoritism to fall to zero.

[63] The quote starting with "In practice, what television's dominance has come to mean is that the inherent value..." is from the widely available excerpt of Al Gore's *The Assault on Reason*, such as the one at www.time.com/time/nation/article/0,8599,1622015,00.html on June 9, 2007. In the book on page 8 the sentence begins with "The inherent value..." The excerpted version is more useful, so we have used that.

[64] Richard Ackerman's review appeared in the Science Library Pad on May 27, 2007 at scilib.typepad.com/science_library_pad/2007/05/assault_on_reas.html.

[65] The paragraph on low leverage points is from *World Dynamics*, by Jay Forrester, 1971, page 95.

[66] The material on William Osler is from en.wikipedia.org/wiki/Diagnosis and en.wikipedia.org/wiki/William_Osler on January 1, 2007. The image is from www.theatlantic.com/issues/98jan/images/doctor.gif, with "The Nation" changed to "The System."

[67] The quote about *Silent Spring* is from Al Gore's introduction to the 1994 edition, p xv.

[68] The quote on corporate sponsored think tanks is from *Global Spin: The Corporate Assault on Environmentalism*, by Sharon Beder, 2002, pages 91 to 93.

[69] The defenders of business-as-usual quote is from *Red Sky at Morning*, by James Speth, 2004, page 6.

[70] The passage on the wine tax in California is from *Gangs of America: The Rise of Corporate Power and the Disabling of Democracy*, by Ted Nace, 2003, page 152.

[71] The quote about Montana and the Anaconda Copper Company is from *Gangs of America: The Rise of Corporate Power and the Disabling of Democracy*, by Ted Nace, 2003, page 154. The short quote on the great shutdown and the date is from www.butteamerica.com/labor.htm on June 9, 2007.

[72] It was Dan Proctor who pointed out to me in December 2006 that Forrester had also addressed the social side of the problem in the second edition of *World Dynamics*.

[73] For example, continuous improvement of a formal process is so productive it lies at the very center of the success of the Japanese post war economic miracle. After World War II,

Japan's automotive, electronics, consumer goods, and other industries took Dr. Deming's teachings to heart, and created the Kaizen ethic. According to *The Elegant Solution: Toyota's Formula for Mastering Innovation*, by Matthew May, 2007, page 167:

"**Kaizen** (ky-zen), the Japanese word for continuous improvement, is all about idea submission, not acceptance. The *de facto* incubator for consistent business innovation, it's the practice that fosters a strong ethos of lab like curiosity in companies like Toyota. And it's a proven way to harvest grassroots productivity.

"Kaizen has three steps: First, create a standard [process]. Second, follow it. Third, find a better way. Repeat endlessly. **Trying to improve and innovate without a standard is like a journey with no starting point.** It's like trying to hit golf balls in the fog.

"The question becomes how to create a standard, which begs the question of what defines a good one. Whether it's a pilot's preflight checklist, a surgeon's protocol, or an autoworker's guide to drive train assembly, there are two criteria:

"A. Clarity – Assume an untrained eye will read it. Make it bulletproof, specific, and complete, to capture the knowledge.

"B. Consensus – Everyone who will employ the standard must agree on it. That forces a shared investigation to ensure that the standard represents the best known method or practice at that specific point in time. The activity in turn facilitates understanding."

Notice how productivity is not a criteria. That's because the important thing is to start with anything as the standard. It is then relentlessly improved, one innovation at a time.

[74] The DNA image is from www.scq.ubc.ca/a-monks-flourishing-garden-the-basics-of-molecular-biology-explained/.

[75] The quote about how long it took Forrester to build the World2 model is from www.std.com/~awolpert/gtr362.html. The title of this HTML page, *General Theory of Religion*, is in error. It should be *System Dynamics Listserv Discussion*.

[76] See www.chelseagreen.com/2004/items/limitspaper/ForTheMedia which says of *The Limits to Growth*, "The book became a bestseller with over 30 million of copies sold in more than 30 translations." I've been unable to find my source for *Silent Spring*.

[77] We have deliberately refrained from speculating what a permanent race to the top would lead to in any detail. This is what futurists do. The trap is they then try to work backward to how to get there from here. This is usually unproductive because this approach fails to recognize that the behavior of complex social systems is an emergent property. It cannot be predicted by inspection of the parts, which is what futurists try to create as they work backward. The chasm of emergence cannot be jumped by going backwards. This explains the poor record futurists have on accurate predictions or being able to make their visions come true.

Better is to start with structural analysis of the present system and then engineer it forward. If the dominant loops, modes, and memetic agents are properly designed and the right high leverage points are used, then the desired emergent properties will appear, along with a number of unanticipated but pleasant surprises.

[78] For a sobering look at how democracy has failed to solve the economic inequality problem, see Thomas Piketty's *Capital in the Twenty-First Century*. Translated from the French in 2014, Piketty has assembled a stunning data set. This shows that starting around

1980, income inequality has returned to levels so high they threaten the social stability of modern democracies. "The consequences for the long-term dynamics of wealth distribution are potentially terrifying, especially when one adds that the return on capital varies directly with the size of the initial stake and that the divergence in the wealth distribution is occurring on a global scale." (p571) On the next page Piketty presents his solution, "a progressive annual tax on capital." But how can this solution or similar ones ever be implemented, if systemic change resistance remains high?

[79] The definition of memetics is from the About page at www.jom-emit.org on June 8, 2007.

[80] The flawed mental model example is from *Business Dynamics: Systems Thinking and Modeling for a Complex World*, by John Sterman, 2000, page 18.

Index

www.ingramcontent.com/pod-product-compliance
Lightning Source LLC
Chambersburg PA
CBHW061405280526
45784CB00001B/372